PREPARATIVE
INORGANIC
REACTIONS

Volume 1

PREPARATIVE

INORGANIC

REACTIONS

Volume 1

Editor

WILLIAM L. JOLLY

Department of Chemistry
University of California
Berkeley, California

1964

INTERSCIENCE PUBLISHERS

a division of John Wiley & Sons

New York · London · Sydney

PREFACE

The most important information that can be obtained from a chemist who is expert in the preparation of a particular class of compound is the theory, or rationale, behind the preparative methods. The non-expert may possibly use this information in three different ways. He may apply the synthetic principles to the preparation of compounds of the same class which have not yet been prepared. He may apply the principles to the preparation of an entirely different class of compound. If he is imaginative enough, he may, by extending or improving on the expert's ideas, develop a new principle that leads the way to new kinds of syntheses. In a highly developed area of chemistry, the synthetic principles are well enough established so that it is relatively easy to extend them to new compounds. It is difficult to improve on the principles and methods. On the other hand, in a new, undeveloped field of chemistry, the synthetic principles are little more than working hypotheses, to be improved upon as experience is gained.

Another kind of information that can be obtained from an experienced synthetic chemist is a critical evaluation of the various methods that have been used for preparing particular compounds. Such information helps other chemists choose preparative methods which are best suited to their needs.

The chapters in Preparative Inorganic Reactions have been written to provide information of the types discussed above. I believe the authors of Volume 1 have been highly successful in their aims.

WILLIAM L. JOLLY

v

CONTRIBUTORS TO VOLUME 1

C. C. Addison Department of Chemistry, University of Nottingham, Nottingham, England

J. C. Bailar, Jr. Department of Chemistry and Chemical Engineering, University of Illinois, Urbana, Illinois

G. W. A. Fowles Department of Chemistry, University of Southampton, Southampton, England

Oskar Glemser Anorganisch-Chemisches Institut der Universität, Göttingen, Germany

J. C. Hileman Department of Chemistry, El Camino College, Torrance, California

S. Kirschner Department of Chemistry, Wayne State University, Detroit, Michigan

N. Logan Department of Chemistry, University of Nottingham, Nottingham, England

Alan G. MacDiarmid Department of Chemistry, University of Pennsylvania, Philadelphia, Pennsylvania

D. F. Martin Department of Chemistry and Chemical Engineering, University of Illinois, Urbana, Illinois

Charles E. Messer Department of Chemistry, Tufts University, Medford, Massachusetts

S. M. Williamson Department of Chemistry, University of California, Berkeley, California

CONTENTS

CHAPTER 1

Coordination Polymers

JOHN C. BAILAR, JR.

University of Illinois, Urbana, Illinois

CONTENTS

I. INTRODUCTION

For most people, the word "polymer" has come to denote an organic material, each molecule of which has been formed by the union of a large number of small molecules, and which is plastic, flexible, or elastic. In its original use, however, the term does not describe the properties of the material or its composition, beyond indicating that it is formed by the combination of repeating units. If the word is used in this broad sense, inorganic chemistry contains examples of many kinds of polymers. Asbestos and mica are, respectively, linear and sheet polymers, and feldspar, kaolin, and quartz are three-dimensional polymers. If one extends the definition far enough, even ionic, crystalline materials such as sodium chloride and magnesium oxide can be considered to be polymeric. Justification for such an extension can be found in the fact that these substances show some plasticity, but, in general, it is unwise to make a defi-

1

nition too broad, for it then loses its power to correlate closely related phenomena.

A. Cyano Polymers

Among synthetic coordination compounds, there are many examples of polymeric substances, but in most cases, their polymeric nature has been recognized only in recent times. Palladium(II) chloride is an in-

finite linear polymer, nickel(II) cyanide is a sheet polymer (1,2) and the

heavy metal ferro- and ferricyanides, in which the iron shows a coordination number of 6, are three-dimensional, infinite polymers (3). Each of these substances has interesting and useful properties, but, in general, they do not possess plasticity, flexibility, or elasticity. It seems unlikely that these properties will be found in any high degree in purely inorganic coordination compounds in which the valence bonds are rather rigidly arranged and the groups binding the metal atoms together are relatively small.

It is to be noted that in nickel cyanide, alternate nickel atoms are found in different environments, and it might be expected that they would show different properties. Each nickel atom surrounded by nitrogen atoms is capable of combining with two molecules of ammonia, thus expanding the coordination number of the metal to 6; those surrounded by carbon atoms do not show this property. The ammonia molecules in the ammoniated species lie on either side of the polymer plane, thus, holding parallel planes apart and allowing the formation of a clathrate compound containing benzene (1,2). This approaches the empirical composition $Ni(CN)_2 \cdot NH_3 \cdot C_6H_6$, but in actual practice, the material usually contains less than one mole of benzene per mole of nickel (4). The material is readily prepared by shaking a suspension of nickel cyanide in ammonium hydroxide with benzene.

The three-dimensional polymer Prussian Blue or Turnbull's Blue is important because of its deep blue color, which is the result of interaction absorption. Half of the iron atoms in it are in the 2+ state and the rest in the 3+ state. It approximates the composition KFeFe(CN)₆, and can be made simply by mixing aqueous solutions of iron(III) with hexacyanoferrate(II) (ferrocyanide) or iron(II) with hexacyanoferrate(III) (ferricyanide). Here again, alternate iron atoms are surrounded by carbon atoms and by nitrogen atoms. Mutual oxidation and reduction of the iron atoms doubtless takes place readily, but one would anticipate that those surrounded by nitrogen would tend to remain in the 3+ state and those surrounded by carbon, in the 2+ state, for coordination with carbon tends to stabilize the lower oxidation states of the metals.

B. Hydroxo Polymers

Another great class of coordination polymers consists of the metallic hydroxides, in which the hydroxyl group serves as the link between metal ions. Titration techniques indicate that these compounds usually contain double bridges (5), but single and triple bridges are also known. Polymerization can be stopped at the dimer stage if suitable blocking groups are present (6)

$$[Co(NH_3)_4(H_2O)_2]^{3+} + OH^- \rightarrow [Co(NH_3)_4(H_2O)OH]^{2+} \rightarrow$$

$$[(NH_3)_4Co\diagup\begin{matrix}H\\O\end{matrix}\diagdown Co(NH_3)_4]^{4+}$$

The first step takes place in aqueous solution and the second, when the dry salt is heated to 150°C., at which temperature the water molecule escapes.

In aquated ions containing more than two water molecules, the process of olation can continue to the formation of ions of high molecular weight. This process takes place spontaneously when an aquated metal ion in water solution is treated with a base, and may continue until the product of reaction precipitates from solution. Growth of the polymer can continue even after precipitation, if the material is allowed to stand in the mother liquor, especially at a somewhat elevated temperature. It is also possible to stop the growth of the polymer before it precipitates, and soluble chromium polymers are important in the tanning of leather. The formation of such "olated" polymers is shown schematically:

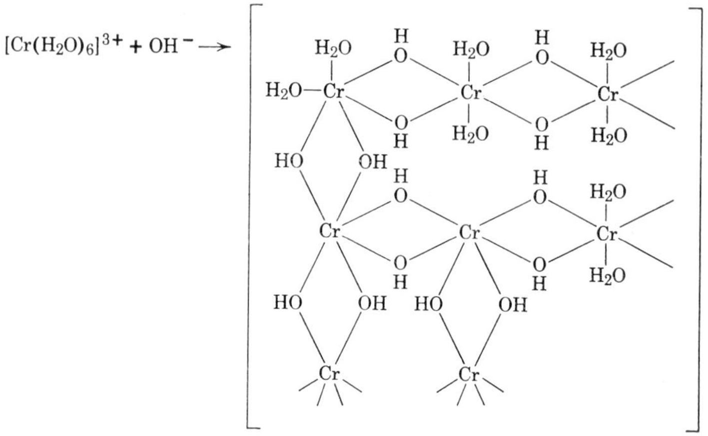

When an "ol" polymer, in water solution or in the form of a sol, is heated, protons are lost from the ol bridges:

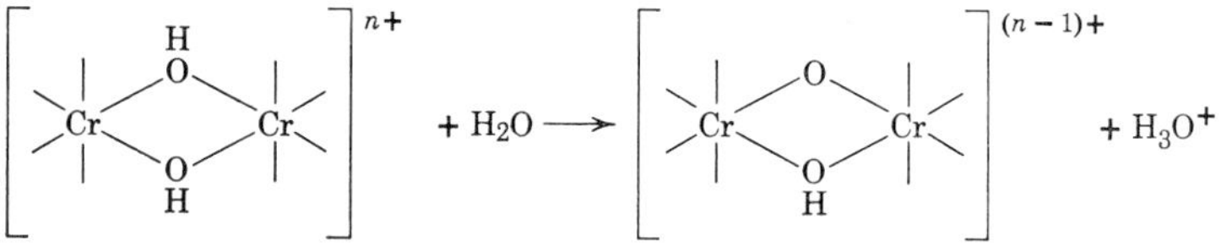

This process, called "oxolation," is only slowly reversed when the temperature is reduced, and the pH of the solution may not return to its original value for several days, the exact time depending upon the nature of the metal ion, the size of the polymer, the degree of oxolation, and other factors. Polymeric structures of this type, both olated and oxolated, can be "broken" by addition of complexing agents which replace the ol or oxol bridges but which cannot form bridges

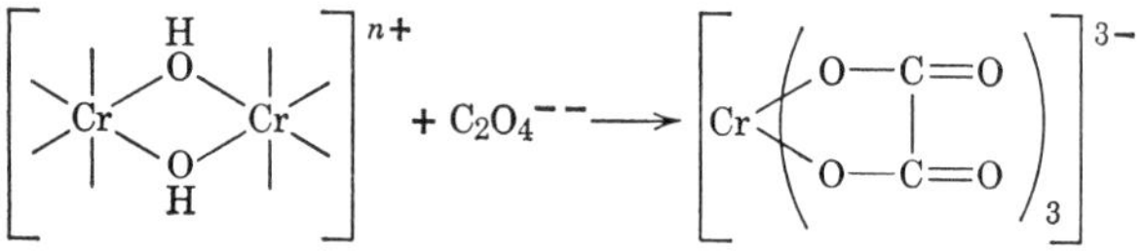

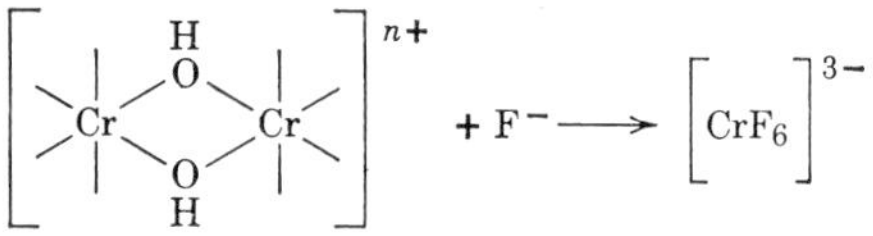

In the first example, bridge formation is prevented by the strong tendency of oxalate ion to form chelate rings; in the second case, by the small tendency of fluoride to share electrons simultaneously with two metal ions (7).

The phenomena of olation and oxolation were extensively studied some years ago by several investigators, notably Thomas, Stiasny, and Gustavson. The subject has been reviewed by Whitehead (8), Rollinson (9), and Pokras (10). Exact preparative techniques can be found in the references given by these reviewers.

The remarkable water repellent "basic chromic stearate" and "basic chromic acrylate" (sold under the trade names Quilon and Volan) and and analogous hydrophobic, oleophobic basic perfluorocarboxylate (Scotch Guard) illustrate an interesting application of the olated polymers. In the monomeric state, these substances are soluble in the lower alcohols. In use, the solution is spread upon the surface to be treated, and the solvent is allowed to evaporate, leaving a residue which adheres tightly to the surface through the action of some of the ligand groups. When the material is heated, olation takes place, binding the chromium atoms together, with hydrocarbon or perfluorocarbon chains protruding from the surface, and protecting it:

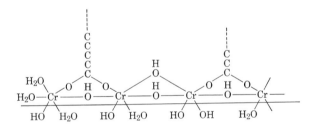

There is good evidence that, in these polymers, the carboxyl groups as well as the hydroxyl groups, serve as bridges.

Schmitz-Dumont (11) has shown that metal ammines can lose protons and ammonia molecules and form polymers which are formally like those generated by olation and oxolation. In these, the bridging members are $-NH_2-$. Directions for the preparation of some polymers of these types are given in refs. (12) and (13).

C. Halo Polymers

The halide ions show a strong tendency to bridge formation and take part in polymerization. As with hydroxide, double-bridge formation is most common, but single and triple bridges are not unknown. In many cases, polymerization is limited to the formation of dimers and trimers, such as

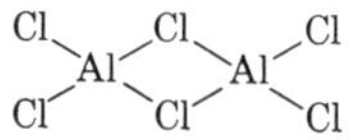

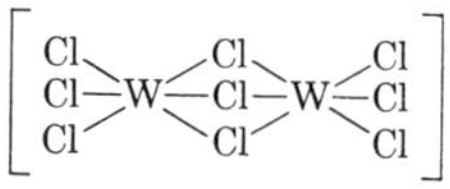

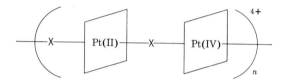

but it may proceed to infinite length, as with $PdCl_2$ and $PtCl_2$. Polymer formation in such cases is spontaneous and, generally, it is difficult or impossible to isolate the monomeric species.

The remarkable singly bridged chains of "trivalent" platinum have been studied by x-ray techniques by Brosset (14) and Cohen and Davidson (15). They contain alternate Pt(II) and Pt(IV) atoms, linked together in infinite chains:

Because of interaction absorption, these polymers are dark in color. They may be prepared by treatment of the appropriate Pt(II) ammine in water solution with one equivalent of halogen, or by grinding together equimolar amounts of the dry platinum(II) tetraammine salt and the dihalo platinum(IV) tetraammine salt:

$$[PtA_4]^{2+} + [PtA_4X_2]^{2+} \rightarrow -[Pt(II)A_4-X-Pt(IV)A_4-X-]^{4+}-$$

Although the fluoride ion shows little tendency to bridge formation, it does so in the remarkable $Sn(II)-F_x-Ni(II)$ complex. Aqueous solutions containing fluoride (in excess) and tin(II) and nickel(II) in widely varying ratios upon electrolysis give an alloy plate containing tin and nickel in approximately equimolar ratios (16). This plate is hard and takes a fine polish; as a result it is being used in England to replace chrome plate. It has been shown (17) that the electrolyte contains a complex containing the two metals in equivalent ratio—the number of bridging fluorides and the complexity of the species are not known. The formation of this ion is particularly interesting in that nickel ion alone does not seem to form a complex with fluoride ion under these conditions (17).

D. Thiooxamido Polymers

Thiooxamide (rubeanic acid) and its N-aliphatic substituted derivatives form an interesting series of linear polymers when the organic material, a divalent transition metal ion, and an equivalent amount of a base are mixed in water solution. The polymers are insoluble in water and precipitate as powders.

$$Me^{2+} + \begin{array}{c} NR\!=\!C\!-\!SH \\ | \\ NR\!=\!C\!-\!SH \end{array} \longrightarrow \cdots Me \begin{array}{c} NR\!=\!C\!-\!S \\ | \\ NR\!=\!C\!-\!S \end{array} Me \begin{array}{c} NR\!=\!C\!-\!S \\ | \\ NR\!=\!C\!-\!S \end{array} \cdots$$

The formation and properties of these polymers have been particularly studied by Hurd and his co-workers at the Mallinckrodt Chemical Works (18). These substances darken upon heating and are of use in duplicating and transfer processes. Some of them (depending upon the nature of the metal and the organic group) are soluble in nonpolar solvents. The rubeanates have been suggested as metal-deactivating agents.

An interesting polymer formed by the action of rubeanic acid and nickel or copper ions has been studied by Amon and Kane (19) who formed it in a layer of an organic plastic. According to their procedure, the sheet of plastic is immersed in a solution of a rubeanate, which penetrates into it, and then into a solution of copper or nickel ions, whereupon the rubeanate polymer forms in the plastic sheet. When the plastic sheet is stretched in one direction, the linear metal–rubeanate polymer orients itself into parallel chains, which allow the sheet to polarize light.

II. PROPERTIES OF COORDINATION POLYMERS

When coordination takes place between a metal ion and a ligand, the properties of both are changed in a variety of ways. It frequently happens that an organic ligand is stabilized towards hydrolysis, chemical reagents, or toward high temperature. For example, hydrolysis of the Schiff base

$$\begin{array}{c} CH_2\!-\!N\!=\!CH\!- \\ | \\ COOH \end{array} \quad \underset{HO}{\bigcirc\!\!\bigcirc}$$

is much more rapid than that of its metal chelates (20)

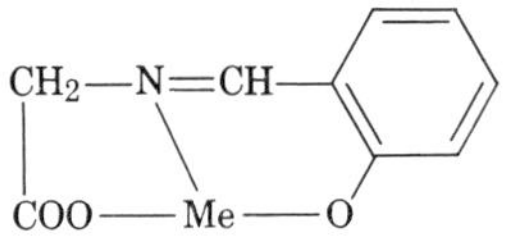

Ethylenediamine is destroyed rapidly by hot, concentrated nitric acid, but when coordinated with cobalt(III), it is not attacked even on heating for many hours. The bis-ketoimine

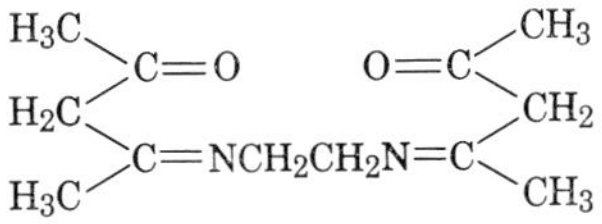

is destroyed by heating to moderate temperatures, but its copper complex

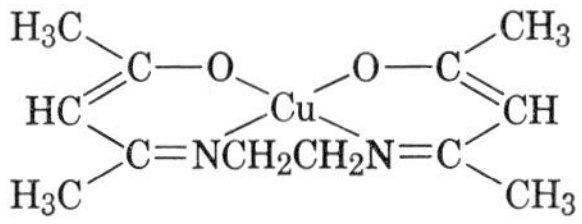

is decomposed only slowly at 380°C. (21). Many attempts have been made during the past decade to incorporate these properties (especially heat stability) into plastic organic materials by inclusion of coordinated metal ions into polymer chains. While none of these attempts has been wholly successful, a great deal has been learned, and there is a good basis for believing that valuable plastics of this sort may be obtained eventually.

III. STRUCTURAL PRINCIPLES

In attempts to prepare coordination polymers, a few basic principles must be kept in mind. These will be mentioned briefly here and will be discussed in more detail in later portions of the paper:

(*1*) Little, if any, plasticity can be expected from the immediate environment of the metal ion, for ordinarily the metal–ligand bonds are rigidly fixed in position. Any flexibility or plasticity which the polymer shows must be contributed by the organic portion of the molecule.

(*2*) The presence of a metal ion can protect and stabilize only the organic material which is close to it, so the organic portion of the polymer must be kept to a minimum. It is evident that this principle and the preceding one work against each other, so some compromise composition must be sought.

(*3*) Thermal, oxidative, and hydrolytic stability are not directly related. For example, one coordination polymer may be stable to heat but not to hydrolysis, while another shows good hydrolytic stability but poor thermal stability. Each polymer must, therefore, be designed for a specific environmental use.

(*4*) Even those coordinate bonds which show the greatest degree of covalency are not purely covalent. This means that rearrangements occur more readily than in most organic polymers. Moreover, the metal ion, even when coordinated, still carries a residual positive charge and the ligand (especially if it be an anion), a residual negative charge. These charges exert interchain attractive forces that tend to give rigidity. In the design of coordination plastics, every effort should be made to employ materials that form bonds which are as nearly purely covalent as possible.

(*5*) Compounds which consist of ions are apt to be less plastic than those which are molecular, for interionic forces are strong enough to detract greatly from plasticity.

(*6*) Coordination polymers which consist of a polymeric cation and a simple anion are generally less stable toward heat than those which are purely molecular. Many of the common anions are themselves reasonably strong coordinating agents and compete with the organic ligand for places in the coordination sphere. At elevated temperatures, the competition may favor the simpler material. The common anions which are not good coordinators (NO_3^-, ClO_4^-, etc.) are strong oxidizing agents and, upon heating, destroy the ligand.

(*7*) Chelation is important in increasing stability, and the usual rules concerning ring size and stereochemical effects apply.

(*8*) The relationship between the coordination number and stereochemistry of a metal ion and the nature of the ligand which is attached to it will determine whether the polymer will be linear, planar, or three dimensional.

(*9*) As in the preparation of purely organic polymers, the reactants must be pure and present in exactly stoichiometric ratios; otherwise, only low molecular weight polymers can be obtained.

(*10*) If a solvent is used for the polymerization reaction, it should be one which is not capable of strong coordination to the metal ion, nor should it be capable of generating units which are strong coordinating agents, e.g., OH⁻ from solvent water.

With a few notable exceptions, the work which has been done on co-ordination polymers has been concerned with linear polymers. To obtain such a substance, a combination such as one of the following is required:

(a) A bis-monodentate coordinating agent that cannot form a chelate ring, and a metal ion with a coordination number of 2.

(b) A bis-bidentate chelating agent in which the two chelating groups do not attach themselves to the same metallic ion, and a metal ion with a coordination number of 4.

(c) A bis-tridentate chelating agent in which the two chelating groups do not attach themselves to the same metallic ion, and a metal ion with a coordination number of 6.

(d) A bis-tetradentate chelating agent in which the two chelating groups attach themselves to different metal ions, and a metal ion with a coordination number of 8.

(e) A bis-bidentate chelating agent as in (b), and a metal ion with co-ordination number of 6, with two of the coordination positions blocked by nonbridging groups.

In each case, in order to satisfy the principles outlined above, the co-ordinating agent and the metal ion must be carefully selected. In many cases, a great portion of the work involved in the study of coordination polymers is concerned with the synthesis of the organic part of the molecule. Of the four possibilities, (b) has been studied far more than any of the others, chiefly because of ease of synthesis. Examples of this type will be described in detail, and examples of some of the others more briefly.

Most of the polymers formed from metals of coordination number 4 contain beryllium, copper, zinc, or nickel ions. Beryllium offers certain advantages, for it is not subject to oxidation or reduction, its co-ordination number seems to be invariably 4, and the beryllium–oxygen bond approaches true covalency. On the other hand, the toxicity of beryllium compounds is a serious drawback. The bis-chelating groups which have been most frequently employed are those of 1,3-diketones, 8-hydroxyquinolines, Schiff bases, phosphinates, and α-amino acid anions. In nearly all cases, the bis-chelating agents have been symmetrical, but this has been chiefly a matter of convenience and ease of synthesis. The two chelating groups may be joined together in any of a great number of ways. With the β-diketones, for example, two general structures have been considered:

where R represents an organic group and Y a connecting unit such as —CH$_2$—, —CH$_2$CH$_2$—, or —C$_6$H$_4$—. In some cases, the two diketone groups of bis-chelates of **a** type have been connected directly to each other, e.g., tetraacetylethane.

IV. SYNTHETIC METHODS

A. 1,3-Diketone Polymers

The first studies of bis-diketone polymers were evidently made by Wilkins and Wittbecker (22) and have been continued by Fernelius and his students (23), Bailar and his students (24), Charles and his collaborators at the Westinghouse Research Laboratories (25), Kluiber and Lewis at the Union Carbide Plastics Co. (26), Drinkard, Ross, and Wiesner (27), and others. In most of the early work, attempts were made to prepare polymers by the same techniques which are used to prepare monomeric complexes of 1,3-diketones. Oh and Bailar (24) say: The method "consists of mixing the ketones and a soluble metal salt in the proper proportions in a water–organic solvent mixture, and adjusting the pH until the chelate precipitates. It is necessary to add an organic solvent such as alcohol, dioxane, or dimethylformamide because of the insolubility of the ligand in water. The success of this method depends upon pH, rate of addition of the base, rate of stirring, temperature and reaction time.

"In some cases polymers were obtained simply by the addition of an excess of an aqueous solution of metal acetate to a dioxane solution of the bis-(β-diketone). The highly insoluble precipitate was purified by extraction with water followed by an organic solvent such as alcohol, acetone, or benzene.

"In the cases in which immediate precipitation did not occur, the pH was adjusted by addition of ammonia or urea. When no precipitation occurred in the reaction mixture even after standing for a long time, the reaction mixture was evaporated slowly in vacuo and the residue was collected. The residue was purified by extraction with water followed by organic solvent." However, on the basis of elementary analysis of the materials which were isolated, Oh and Bailar showed that the polymers which they obtained in this way contained hydroxyl groups, either as ol bridges or as chain ends. These groups tend to make the polymers high melting and insoluble in nonpolar solvents. In a typical case, copper(II) ion was allowed to react with 1,2-diacetyl-1,2-dibenzoylethane. The product of this reaction gave the following analysis: C, 60.94; H, 4.19; Cu, 17.68. The calculated values for $(C_{20}H_{16}O_4)_{13}$-$(OH)_2Cu_{14}$ are C, 61.17; H, 4.16; Cu, 17.48. The material did not melt at 250°C. and was found to be insoluble in nonpolar solvents. It decomposed only slowly at 250°C.

In other cases reported by Oh and Bailar, precipitation took place before polymer formation could begin. With copper(II) ion and 1,1,3,3-tetraacetylpropane, they obtained a substance showing the following analysis: C, 50.28; H, 6.17; Cu, 12.54. The calculated values for $(C_{11}H_{14}O_4)_2Cu \cdot 2H_2O$ are: C, 50.82; H, 6.16; Cu, 12.22. This material softened at 140°C., and was found to be soluble in the common organic solvents. It decomposed fairly rapidly at 250°C. It is interesting that there was little loss in weight below 200°C., so the water molecules must be quite tightly attached.

Charles (25d) prepared the copper(II) complex of tetraacetylethane by adding aqueous copper(II) acetate solutions to dilute solutions of the ligand in tetrahydrofuran. The products were polymeric, consisting of approximately five repeating units ($n = 5$). Elemental and infrared analysis did not indicate the presence of water or hydroxyl groups. These polymers were of about the same heat stability as copper acetylacetonate, but less stable than the ligand itself. The magnesium, nickel(II), and cobalt(II) polymers, however, are of somewhat greater stability (25f).

Another method, which it was hoped would avoid the difficulties mentioned above, consists of heating the metal acetylacetonate with the bisdiketone, and distilling off the liberated acetylacetone in vacuo. Using this method, Oh and Bailar obtained polymers of low molecular weight, i.e. $(C_{17}H_{18}O_4)_4Be_3$, in which the ketone is 2-phenyl-1,1,3,3-tetraacetylpropane. Fernelius (23) prepared beryllium complexes of sebacoyldiacetophenone in this way, obtaining (in different runs) polymers for which the value of n varied from 3.8 to 9.1. He obtained the same polymer ($n = 3.4$) by heating the reactants together in refluxing decalin, from which the liberated acetylacetone was allowed to distill. Kluiber and Lewis (26) have combined the two diketone functions into a single bis-diketone molecule in which the two functional units were connected by chains such as $+CH_2+_x$, $-NH(CH_2)_xNH-$ or $-O(C-$

H$_2$)$_x$O—.   From such tetraketones they obtained cyclic monomers and dimers which, upon heating, rearranged into polymers of high molecular weight.   Those in which the connecting group is +CH$_2$+$_x$ can be schematically shown as follows:

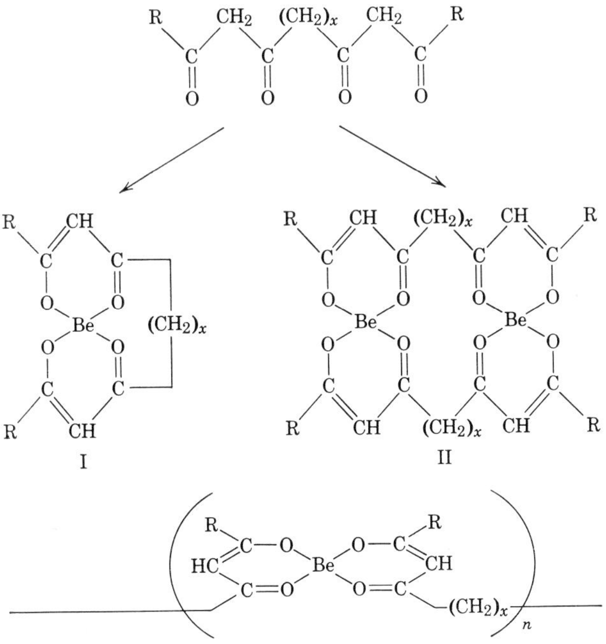

Monomers of type I were formed from bis-diketones in which x is 6 or more, and those of type II, from those in which x is 6 or less.   For the compounds in which x is 6, they give the following directions:

(1) *Preparation of the Cyclic Monomer and Dimer Chelate of 2,4,11,13-Tetraketotetradecane.*   The 2,4,11,13-tetraketotetradecane (2.54 g., 0.01 mole) and 1.77 g. beryllium sulfate tetrahydrate (0.01 mole) were stirred in 50 ml. of water and 10 ml. of dioxane for 1 hr., neutralized with sodium hydroxide to pH 8, and stirred at room temperature for an additional 16 hr.   The crude precipitate was transferred to a sublimation tube and sublimed at 200 °C. (0.2 mm.) for 2 weeks.   The sublimate crystallized on the sides of the tube in two zones.   The lower zone (warmer area) produced 56 mg. of material, m.p. 230 °C. (polymerizes).   The upper zone (cooler area) produced 379 mg. of material which polymerized without observable melting at 110 °C.   This upper zone corresponded to the area in which the cyclic beryllium chelate of 2,4,13,15-tetraketohexadecane condensed and, by analogy and in agreement with infrared and analyt-

14 J. C. BAILAR, JR.

ical data, this material has been assigned the structure of the cyclic monomer. This was confirmed by molecular weight determinations (freezing point depression in benzene) which gave 277 (calcd. 261). The less volatile component had an infrared spectrum essentially identical with the monomer and, in agreement with its analytical data and volatility, has been assigned the dimer structure. Cryoscopic molecular weight determinations in benzene gave 555 (calcd. 522). The recovered dimer melted at 215°C. (polymerized).

(2) *Polymerization of Macrocycles.* Upon heating above their melting points, these crystalline macrocycles spontaneously polymerized producing amorphous (by x-ray) materials which dissolved slowly in benzene. These polymers have intrinsic viscosities in the range 0.1–2.7, characteristic of high molecular weight polymers. This characterization of high molecular weight was confirmed by the mechanical properties of several of these materials. As example, the polymer III, R′ = CH₃, R = (CH₂)₈, intrinsic viscosity 0.9, was flexible at room temperature (glass transition temperature 35°C.) and had a tensile modulus at 25°C. of 122,000 p.s.i. and a tensile strength (100% elongation/min.) of 2710 p.s.i. Melt index flow rates at 198°C. (ASTM Test D 1238–52T) were in the range 0.1–2.0 dg./min. again characteristic of high molecular weight polymers. The analytical and infrared data on these polymers were in agreement with the assigned structures.

Klein and Bailar (24) have prepared beryllium–diketone polymers by an entirely different technique. These were of low molecular weight, but the method should be adaptable to the formation of long chains. Their monomeric units were in which X represents —COOC₂H₅, —OH, or

—NH₂. The monomers containing ester groups were converted to polyesters and polyamides in melt systems, and those containing hydroxy or amino groups were polymerized by reaction with diphenyldichlorosilane, terephthaloyl chloride, and diisocyanates. The polyesters, polyamides, and some of the polymers prepared from the hydroxy monomer were glasses and were found to be soluble in nonpolar solvents. The remaining polymers were insoluble powders. None of the substances prepared in this study were stable above 200°C.

A somewhat similar plan was adopted by Patterson and his co-workers (28), who introduced another interesting feature in using a 6-coordinate metal, two of the coordination positions of which were blocked by a molecule of a diketone. Their monomer was

This was polymerized by reactions with diols. The resulting materials had molecular weights of 940–2600, were soluble in benzene, and could be drawn from a melt in long fibers. None of them, however, showed stability about 200°C.

Berlin and Matveeva (29) likewise used aluminum with two coordination positions blocked by a molecule of a diketone. When the monomer

was heated with any of a variety of bis-diketones, isopropyl alcohol escaped and polymerization took place. The resulting materials were of low molecular weight.

B. 8-Hydroxyquinoline Polymers

The use of bis-8-hydroxyquinolines as precipitants in analysis has long been practiced (30), the insolubility of the precipitates doubtless being greatly emphasized by their polymeric nature. The study of metal complexes of bis-8-hydroxyquinolines as plastic polymers has recently been undertaken by several investigators (24,31,32). As with the bis-diketones, polymerization has been effected by precipitation from water solution, from melts, and by interfacial techniques. The materials obtained by precipitation reactions are probably of low molecular weight because of their great insolubility and rapid formation. Judd (24) was able, in one case, to get needle-like crystals by extremely slow mixing of the metal ion and ligand (by diffusion), but there was no evidence of plasticity or flexibility in these crystals.

Horowitz and Perros (31) used both thermal and solution methods for the preparation of bis-quilolinol polymers. The following directions are quoted from their paper:

(1) Thermal Polymerization. The coordination polymers were pre-pared in vacuum by heating stoichiometric amounts of the ligand and the metal acetylacetonate at 290°C. The reaction vessel and the jar containing the electrically heated silicone oil were made of glass so that it was possible to observe the entire reaction. Once formed, the polymer was quickly removed from the temperature bath after a minimum time had elapsed, thus, minimizing the possibility of decomposition. By re-ducing the reaction time from hours to minutes, the problems encoun-tered by other workers in preparing coordination polymers of bis-oxine derivatives by thermal polymerization were largely eliminated. For example, Korshak and co-workers (32), using a heating period of 3 hr., reported the preparation of the dimer and trimer of the nickel(II) and copper(II) bis-(8-hydroxy-5-quinolyl)-methane polymers, respectively. Berg and Alam (30) reported that they were unsuccessful in preparing the zinc(II) 8,8'-dihydroxy-5,5'-biquinolyl polymer by thermal polymeriza-tion.

The chelates of acetylacetonate were used in these reactions because Von Hoene, Charles, and Hickam (25b) reported that these compounds, when heated in vacuum, decomposed to give volatile products consisting of acetylacetone, acetone, acetic acid, methane, carbon monoxide, and carbon dioxide. Thus, in thermal polymerizations in vacuum, the by-products of the reaction could be removed while the metal would be left behind to coordinate with the bis-ligand.

Prior to each run, the reaction vessel was first flushed with nitrogen that had been purged of oxygen. Then the reaction vessel containing the reactants was evacuated to approximately 10^{-3} mm. Hg and was im-mersed in the silicone oil bath. After all visible signs of reaction ceased, the product was removed, either as a free-flowing powder or as clumps, wrapped in filter paper for insertion into a paper extraction thimble, and extracted overnight with reagent-grade dimethylformamide, using a Soxhlet apparatus. The polymer itself was essentially insoluble in the hot solvent, but unreacted ligand, metal acetylacetonate, and other soluble materials were removed during the extraction. The solution in the extraction flask was usually highly colored, and precipitated solid material was usually present. The apparatus was arranged to prevent any solid material from being physically swept into the flask. The polymer was then extracted with absolute alcohol for several hours to re-move the last trace of free dimethylformamide and any nonpolymeric alcohol-soluble constituents. The polymers showed no evidence of dis-solving in alcohol although some of the extracts were colored. The ex-tracted polymers were normally dried in vacuum over P_2O_5 at 140°C. although a few polymers were also redried at 190°C. Based on elemental analysis there was no evidence that the polymers suffered any decomposition at these temperatures.

(2) *Polymerization in Solution.* The hydrated acetate of the divalent metal, dissolved in dimethylformamide, was slowly added to a stirred dimethylformamide solution containing the ligand and maintained at 120°C. In all cases precipitation occurred almost immediately upon mixing the two reactants. The contents of the reaction flask were allowed to cool overnight at room temperature, and the product was collected, extracted, and dried in the manner previously described.

Horowitz and Perros made a careful study of the factors that influence the thermal stability of polymers of bis-(8-hydroxy-5-quinolyl)-methane and have found it to be closely related to such periodic properties of the metal as ionic potential and electronegativity. Some of the polymers which they prepared were quite stable at 500°C. *in vacuo.* This is comparable to the decomposition temperatures found by Charles and his co-workers for monomeric 8-quinolinol chelates (33). In the case of polymers prepared by coordination of metal ions with bis-Schiff bases, it has been observed that those in which the linking group is —SO$_2$— are very much more stable toward heat than those in which it is —CH$_2$—. If the same relationships hold for the quinolinol complexes, it should be possible to prepare bis-quinolinol polymers of very great thermal stability.

C. Phosphinate Polymers

Block and his co-workers at the Pennsalt Chemicals Corporation (34) have shown that the diphenylphosphinate ion can serve as a bridging group between two metal ions, and can thus lead to polymer formation. They first reported the reactions of diphenylphosphinic acid with beryllium acetylacetonate and chromium(III) acetylacetonate, which in each case led to a dimer in which two diphenylphosphinate groups served to bind the metal ions together. The chromium complex has the structure

In their second article on this subject, they showed that polymeric substances can be formed by replacing a second molecule of acetylacetone from each chromium(III) acetylacetonate molecule by diphenylphosphinate. They say, "Although we have produced such products by several techniques, the most satisfactory has been the direct reaction between chromium(III) acetylacetonate and diphenylphosphinic acid at temperatures from 175 to 250°C. under a slow sweep of nitrogen. Heating is continued until acetylacetone can no longer be detected in the exit gas by reaction with ferric ion. The residue then is separated into fractions by extracting successively with ethanol, benzene, and chloroform in a Soxhlet extractor. After these extractions the amount of insoluble residue from the reaction between 10.5 g. $Cr(AcCHAc)_3$ and 13.1 g. $Ph_2P(O)OH$ was 2.7 g. at 175°C., 5.8 g. at 200°C., and 9.4 g. at 250°C., a marked increase with increasing temperature. A fraction insoluble in ethanol but soluble in benzene contained 9.4% Cr, 9.3% P, 58.5% C, and 4.9% H; a fraction insoluble in ethanol and benzene but soluble in chloroform contained 8.6% Cr, 10.8% P, 59.6% C, and 4.8% H; a fraction insoluble in all three contained 8.6% Cr, 10.3% P, 59.1% C, and 4.9% H. The latter two agree well with 8.88% Cr, 10.58% P, 59.49% C, and 4.65% H, the values calculated for $[Cr(AcCHAc)(OPPh_2O)_2]_n$; the former with the values 9.66% Cr, 9.21% P, 58.46% C, and 4.83% H calculated for $(AcCHAc)[Cr(AcCHAc)(OPPh_2O)_2]_4Cr(AcCHAc)_2$.

"Ebullioscopic measurements in benzene have yielded number-average molecular weights of 1940 to 2633 for benzene–soluble fractions and up to 10,870 for chloroform-soluble fractions. The benzene-soluble fraction thus contains, on the average, four to five chromium-containing units per polymer segment assuming either acetylacetonate or diphenylphosphinate end groups, the chloroform-soluble, about eighteen. We have not been able to determine the molecular weight of the insoluble fractions yet, but, since some of the soluble as well as the insoluble fractions have compositions agreeing with the infinite polymer, the difference between them presumably is one of molecular weight, the less soluble fractions having higher molecular weights."

Block and his co-workers suggest the following structure for these polymers:

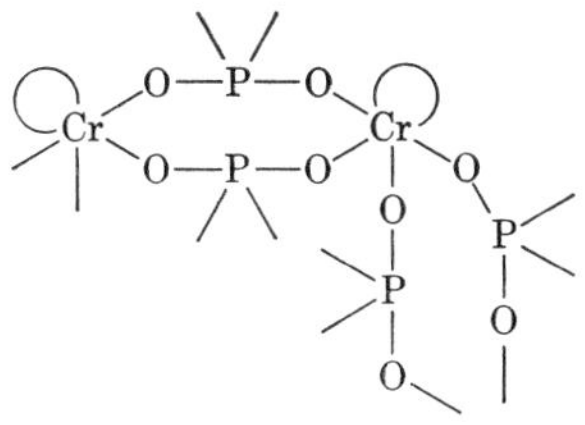

and in view of the isolation of the dimer, there seems little doubt that this is correct.

In the third paper Block reports the formation of polymers of good heat stability from four-coordinate tetrahedral divalent metal ions such as Be^{2+} and Zn^{2+} and diphenylphosphinate. The following directions and data are quoted from his paper:

"We have prepared polymeric metal phosphinates in a variety of ways; however, at this time only a typical mode of preparation will be given for each example. $[Be(OPPh_2O)_2]_x$ was prepared by heating an intimate mixture of 5.0 g. $Be(AcCHAc)_2$ and 21.0 g. $Ph_2P(O)OH$ to 100°C. in a microdistillation apparatus at 1 mm. pressure for 3.5 hr. The mixture was cooled, reground, and reheated under the same conditions two more times, yielding, after extensive extraction with EtOH, 8.6 g. of a product containing 2.25% Be, 14.09% P, 65.43% C, and 4.60% H; calcd. for $Be(OPPh_2O)_2$: 2.03% Be, 13.91% P, 65.01% C, and 4.55% H. $[Zn(OPPh_2O)_2]_x$ was prepared through interfacial polymerization of a 100 ml. aqueous solution containing 2.195 g. $Zn(OAc)_2 \cdot 2H_2O$ with a 200-ml. benzene solution of 4.36 g. $Ph_2P(O)OH$ in a Waring Blendor. The mixture was agitated for an hour, then filtered and washed extensively with ethanol. The dried product weighed 4.1 g. and contained 13.4% Zn, 12.14% P, 57.61% C, and 4.18% H, as compared with the calcd. values for $Zn(OPPh_2O)_2$ of 13.08% Zn, 12.40% P, 57.68% C, and 4.03% H. $[Zn[OP(Ph)(Me)O]_2]_x$ resulted from the reaction of 1.56 g. $Ph(Me)P(O)OH$ with 1.10 g. $Zn(OAc)_2 \cdot 2H_2O$ in 300 ml. of absolute ethanol. The mixture was filtered after 1 hr. of stirring at room temperature to yield, after three ethanol washings, 1.12 g. of a white precipitate containing 17.7% Zn, 16.14% P, 44.88% C, and 4.56% H; calcd. for $Zn(OP(Ph)(Me)O)_2$: 17.40% Zn, 16.49% P, 44.77% C, and 4.29% H. A number average molecular weight of 5600 (15 units) was obtained by ebulliometry in benzene. The white solid turns into a glass at approximately 150°C. and softens at about 200°C. Long flexible fibers can be pulled from the melt or from the product when wet with benzene at room temperature.

"The structure of these polymers probably is related to that which we postulated for $[Cr(AcCHAc)(OPPh_2O)_2]_x$, with double phosphinate bridges between metal atoms, i.e.,

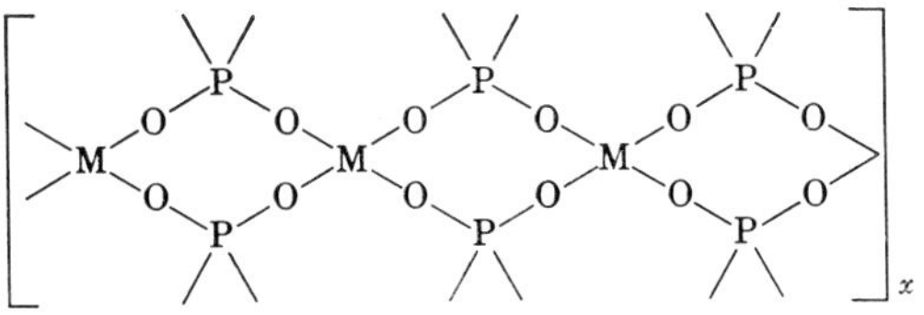

D. Organic Polymers as Ligands

"Basic beryllium acetate," $Be_4O(C_2H_3O_2)_6$, is an extremely stable complex in which the four beryllium atoms are bound to the central oxygen atom, and surround it tetrahedrally. The carboxyl group of an acetate ion bridges each of the six edges of the tetrahedron. Marvel and M. M. Martin (35) prepared an analogous complex in which two-thirds of the acetate ions were replaced by succinate. This formed a plastic polymer which was stable to 450°C.; unfortunately, it disproportionated upon standing, evidently being converted to the basic acetate and the highly cross-linked basic succinate.

Lions and K. V. Martin (36) and Marvel and Tarkoy (37) proposed the formation of coordination polymers by the metallation of preformed organic polymers containing suitably placed donor atoms. Both pairs of workers used Schiff bases, Lions and Martin employing the reaction

and Marvel and Tarkoy, the reaction

Lions reported that the polymer prepared from benzidine did not coordinate well, but the ones made from hexamethylenediamine and ethylenediamine reacted readily with iron(II) sulfate to give complexes which were ferromagnetic, insoluble in the usual organic solvents, and stable to 300°C. They are postulated to have the structure

Exact directions for the preparation are given in the original paper.

Marvel and Tarkoy (37) found that polymers in which the connecting unit Y was —CH$_2$— did not have good heat stability, but those in which Y represented —SO$_2$— could stand reasonably high temperatures.

Goodwin and Bailar (38) used the same bis-salicylaldehyde, but formed Schiff bases with triethylenetetramine and diethylenetriamine. The first of these gave a coordination unit that filled all six positions on six coordinate ions (Co^{3+}, Fe^{3+}, and Al^{3+}). Since the ligand unit lost two protons upon coordination and thus was an ion of double negative charge, each unit of the metallated polymer had a charge of 1+, which was balanced by a simple negative ion (a) as shown on page 22. This feature, as mentioned earlier, leads to thermal instability. The polymers made from diethylenetriamine were thermally more stable, for the added negative group was used to fill a coordination position (b). In this latter case, when the negative group was replaced by ammonia and became an anion, the stability fell again (c)

All of these polymers were insoluble in nonpolar solvents and decomposed before melting. The chief difficulty encountered in their preparation lay in the fact that the original organic polymers were soluble in nonpolar solvents, but not in water, whereas with the metal ions, the reverse was true. The polymer formation, therefore, always took place in a two-phase system and it was difficult to achieve coordination at *every* available spot.   Since the metal is present to lend stability, it is essential that coordination be complete.   It may be possible, with further work, to find a metal complex which will dissolve in a nonpolar solvent, and from which the organic polymer will replace the oil-solubilizing groups.

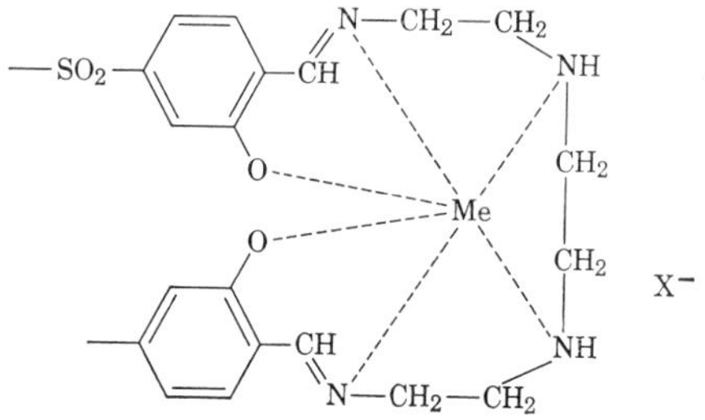

a

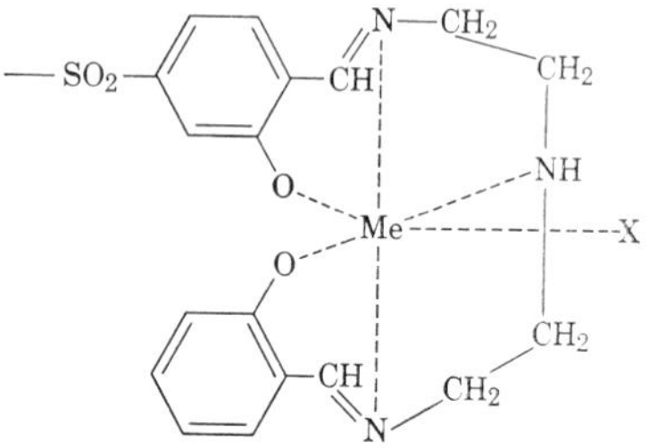

b

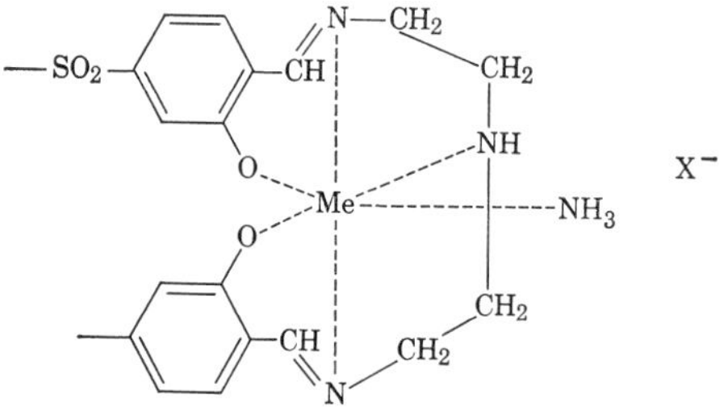

c

E. Polymers from Carbonyls

Goan and his associates at Melpar, Inc. (39) have recently observed that compounds containing active hydrogen will react with metal carbonyls under the influence of ultraviolet light:

$$M(CO)_n + 3LH \xrightarrow{h\nu} ML_3 + nCO + {}^3/_2H_2$$

(for a metal with a coordination number of 6).   They are working on this as a possible means of forming coordination polymers, and have carried out experiments with diphenylphosphinic acid, bis-β-diketones, and bis-8-hydroxyquinolines (40).   Iron pentacarbonyl and chromium hexacarbonyl react readily with diphenylphosphinic acid in tetrahydro-

furan or toluene, the products from the chromium carbonyl being somewhat soluble in toluene. The structures seem to be

The following procedure is illustrative: A toluene solution of chromium
hexacarbonyl and diphenylphosphinic acid, in 1:2 ratio, was refluxed
and exposed to ultraviolet light for 72 hr.; a pale-green solid separated,
leaving a dark-green filtrate. The solid gave fractions which showed
molecular weights of 4670, 6500, and 7130. The fraction of highest
molecular weight was stable to 360°C.

F. Phthalocyanine Polymers

The very great stability of metallophthalocyanines has led to the hope
that polymeric substances of this sort could be formed by the use of
pyromellitic acid or its derivatives in place of phthallic acid. This
should lead to a planar polymer. Several investigators have studied
the problem. Drinkard and Bailar (41) used the same procedure that is
used for making simple phthalocyanines, and obtained polymers consisting of only a few units. The great insolubility of the material probably prevented further growth. From elementary analysis and titration of the peripheral carboxyl groups, it was possible to estimate the
size and shape of the polymer as shown on page 24.

For some unexplained reason, the polymer sheet did not group equally
in both dimensions, but became elongated.

Drinkard and Bailar give the following directions for a typical run:
A mixture of 20 g. pyromellitic dianhydride ($0.092M$), 8 g. anhydrous
copper(II) chloride ($0.06M$), 108 g. urea ($1.8M$), and a catalytic amount
of ammonium molybdate was heated at 160°C. for 30 min. The product
was washed with $6N$ hydrochloric acid, dissolved in 200 ml. concentrated
sulfuric acid, and reprecipitated by dilution with 3 liters of water. The
precipitate was washed by decantation with 24 liters of water in 3 liter
portions, filtered, and air-dried. The product was blue-green.

Anal. Found: C, 54.01; H, 1.60; N, 14.25; equiv. wt., 133.

Marvel and M. M. Martin (42) performed very similar experiments, using a mixture of phthallic anhydride and pyromelitic anhydride to insure the formation of a linear chain. Their polymers consisted of only a few units. Both groups of workers observed that the thermal stability of these polymers is much less than that of the monomeric copper phthalocyanine. If some way of obtaining large molecular weights can be found, these polymers may still serve a useful purpose because of the high degree of conjugation in the system, and the possibility of semiconduction.

Other workers on phthocyanine polymers (43) have used other starting materials (nitriles rather than acids or anhydrides) and have followed slightly different procedures, but the results have not been significantly different.

A similar type of polymer has been prepared by Berlin, Matveeva, and Sherle (44) using tetracyanoethylene as the organic chelater. They heated a mixture of one mole of copper(II) acetylacetonate and two moles of tetracyanoethylene to 160–300°C. in a vacuum, whereupon acetylacetone escaped, leaving an infusible black polymer which was insoluble in organic solvents, alkalis, and dilute acids. The infrared spectrum indicates that the structure is

REFERENCES

1. Powell, H. M., and J. H. Rayner, *Nature* 163, 566 (1949).
2. Rayner, J. H., and H. M. Powell, *J. Chem. Soc.* 1952, 319.
3. Keggin, J. F., and F. D. Miles, *Nature* 137, 577 (1936).
4. Drago, R. S., personal communication.
5. See, for example, M. Kilpatrick and L. Pokras, *J. Electrochem. Soc.* 100, 85 (1953) and L. G. Sillén, *Proceedings of the Symposium on Coordination Chemistry*, Danish Chemical Society, 1954, p. 74.
6. Werner, A., *Chem. Ber.* 40, 284 (1907).
7. Marion, S. P., and A. W. Thomas, *J. Colloid Sci.* 1, 221 (1946).
8. Whitehead, T. H., *Chem. Rev.* 21, 113 (1937).
9. Rollinson, C. L., in *The Chemistry of Coordination Compounds*, edited by J. C. Bailar, Jr., Reinhold, New York, 1956, Chap. 13.
10. Pokras, L., *J. Chem. Educ.* 33, 152, 223, 282 (1956).
11. For a review of this work, see O. Schmitz-Dumont "Hochpolymere Koordinationsverbindungen von Übergangselementen," a chapter in *Inorganic Polymers*, An International Symposium, The Chemical Society, Special Publication No. 15, London, 1961. This article gives references to Schmitz-Dumont's experimental work.
12. Schmitz-Dumont, O., and F. Raabe, *Z. Anorg. Allgem. Chem.* 277, 297 (1954).
13. Schmitz-Dumont, O., and H. Schulte, *Z. Anorg. Allgem. Chem.* 282, 253 (1955).
14. Brosset, C., *Arkiv Kemi, Mineral, Geol.* 25A, No. 19 (1948).
15. Cohen, A., and N. Davidson, *J. Am. Chem. Soc.* 73, 1955 (1951).
16. Cuthbertson, J. W., N. Parkinson and H. P. Rooksby, *J. Electrochem. Soc.* 100, 107 (1953).
17. Rau, R. L., and J. C. Bailar, Jr., *J. Electrochem. Soc.* 107, 745 (1960).
18. Hurd, R.N., G. DeLaMater, G. C. McElheny, R. J. Turner, and V. H. Wallingford, *J. Org. Chem.* 26, 3980 (1961).
19. Amon, Jr., W. F., and M. W. Kane (to Polaroid Corp.), U. S. Patent 2,505,085 (April 25, 1950).

20. Eichhorn, G. L., and N. D. Marchand, *J. Am. Chem. Soc.* **78**, 2688 (1956).
21. Interrante, L. V., personal communication. A. Combes, *Compt. Rend.* **108**, 1252 (1889) and G. T. Morgan and J. D. M. Smith, *J. Chem. Soc.* **1926**, 912, indicate that the substance is decomposed slowly at red heat, but this seems to be overly optimistic.
22. Wilkins, J. P., and E. L. Wittbecker (to E. I. du Pont de Nemours Co.), U. S. Patent 2,659,711 (Nov. 17, 1953).
23. Fernelius, W. C., Wright Air Development Center Report W.A.D.C. 56-203, Part I, Oct. 1956; Part II, Sept. 1957; Part III, Sept. 1958.
24. Bailar, Jr., J. C., W. C. Drinkard, and M. L. Judd, W.A.D.C. Technical Report 57-391, Sept. 1957; J. C. Bailar, Jr., J. McLean, and M. L. Judd, W.A.D.C. Technical Report 58-51, April 1958; J. C. Bailar, Jr., K. V. Martin, M. L. Judd, and J. McLean, W.A.D.C. Technical Report 57-391, Part II, August 1958; J. C. Bailar, Jr., M. L. Judd, and J. McLean, W.A.D.C. Technical Report 58-51, Part II, May 1959; J. C. Bailar, Jr., W.A.D.C. Technical Report 59-427, Jan. 1960; J. C. Bailar, Jr., H. A. Goodwin, M. Moraghan, J. McLean, C. Fujikawa, and L.-C. Chen, W.A.D.C. Technical Report 58-51, Part III, April 1960; R. Klein and J. C. Bailar, Jr., in press; J. S. Oh and J. C. Bailar, Jr., *J. Inorg. Nucl. Chem.* **24**, 1225 (1962).
25a. Charles, R. G., and M. A. Pawlikowski, *J. Phys. Chem.* **62**, 440 (1958); b. J. Von Hoene, R. G. Charles, and W. M. Hickam, *ibid.* **62**, 1098 (1958); c. R. G. Charles, W. M. Hickam, and J. Von Hoene, *ibid.* **63**, 2084 (1959); d. R. G. Charles, *ibid.* **64**, 1747 (1960), **65**, 568 (1961); e. R. G. Charles, *J. Polymer Sci.* **A1**, 267 (1963); f. R. G. Charles, *J. Inorg. Nucl. Chem.* **25**, 45 (1963).
26. Kluiber, R. W., and J. W. Lewis, *J. Am. Chem. Soc.* **82**, 5777 (1960).
27. Drinkard, W. C., D. Ross and J. Wiesner, *J. Org. Chem.* **26**, 619 (1961).
28. Patterson, T. R., F. J. Pavlik, A. A. Baldoni, and R. L. Frank, *J. Am. Chem. Soc.* **81**, 4213 (1959).
29. Berlin, A. A., and N. G. Matveeva, *Russ. Chem. Rev.* **29**, 119 (1960).
30. For a recent example, see E. W. Berg and A. Alam, *Anal. Chim. Acta* **27**, 454 (1962).
31. Horowitz, E., and T. P. Perros, *J. Inorg. Nucl. Chem.* **26**, 139 (1964)
32. Korshak, V. V., S. V. Vinogradova, and T. M. Babchinitser, *Polymer Science* (*USSR*)(*Engl. Transl.*) **2**, No. 4, 344 (1960).
33. Charles, R. G., and A. Langer, *J. Phys. Chem.* **63**, 603 (1959); R. G. Charles, *J. Inorg. Nucl. Chem.* **20**, 211 (1961); *Anal. Chim. Acta* **27**, 474 (1962); R. G. Charles, A. Perrotto, and M. A. Dolan, *J. Inorg. Nucl. Chem.* **25**, 45 (1963); Westinghouse Research Laboratories, Scientific Research Paper 123-4000-P7 (Nov. 16, 1961).
34. Block, B. P., E. S. Roth, C. W. Schaumann, and L. R. Ocone, *Inorg. Chem.* **1**, 860 (1962); B. P. Block, J. Simkin, and L. R. Ocone, *J. Am. Chem. Soc.* **84**, 1749 (1962); B. P. Block, S. H. Rose, C. W. Schaumann, E. S. Roth, and Joseph Simkin, *J. Am. Chem. Soc.* **84**, 3200 (1962).
35. Marvel, C. S., and M. M. Martin, *J. Am. Chem. Soc.* **80**, 619 (1958).
36. Lions, F., and K. V. Martin, *J. Am. Chem. Soc.* **79**, 2733 (1957).
37. Marvel, C. S., and N. Tarkoy. *J. Am. Chem. Soc.* **79**, 6000 (1957).
38. Goodwin, H. A., and J. C. Bailar, Jr., *J. Am. Chem. Soc.* **83**, 2467 (1961).
39. Goan, J. C., C. H. Heuther, and H. E. Podall, in press.
40. Podall, H. E., and T. L. Iapalucci, *J. Polymer Sci.* **B1**, 457 (1963).

41. Drinkard, W. C., and J. C. Bailar, Jr., *J. Am. Chem. Soc.* **81,** 4795 (1959).
42. Marvel, C. S., and M. M. Martin, *J. Am. Chem. Soc.* **80,** 6600 (1958).
43. Sprague Electric Co. Final Report, Contract No. DA 36-039-SC-87 (U. S. Army Signal Corp.) 1952; E. A. Lawton and D. D. McRitchie, W.A.D.C. Technical Report 57-642, November 1957.
44. Berlin, A. A., N. G. Matveeva, and A. I. Sherle, *Izv. Akad. Nauk. SSSR; Otd. Khim. Nauk* **1959,** 2261.

CHAPTER 2

Optically Active Coordination Compounds

STANLEY KIRSCHNER

Wayne State University, Detroit, Michigan

CONTENTS

I. INTRODUCTION

The demonstration of optical activity in a coordination compound via the resolution of a racemic mixture of its enantiomers or by other techniques is often of considerable importance in the determination of its structure. Since optically active complexes synthesized in the laboratory almost without exception form as *racemic* modifications containing both *dextro* and *levo* isomers and not as just a single enantiomer of one configuration, techniques have had to be developed to resolve the racemic modifications into their component enantiomers as a means of demonstrating asymmetry (or dissymmetry) and as a means of providing additional information about the complexes, such as the nature and mechanism of their reactions, their absolute configuration, their composition,

29

etc. Additionally, techniques have been developed to obtain a single enantiomer by more direct means without the utilization of resolution procedures.

An early example of the importance of resolving racemic modifications of asymmetric complexes was the resolution carried out by Werner (1) of the *racemic* tris(ethylenediamine)cobalt(III) cation, DL-[Co(en)₃]³⁺ (1).

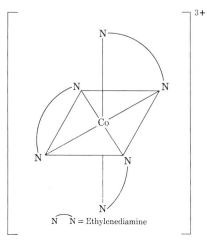

N⏜N = Ethylenediamine

1

The demonstration of optical activity in this complex ion confirmed the octahedral configuration proposed for it (1) and ruled out the planar-hexagonal and trigonal prism structures as possible configurations of this ion, since they would contain a plane of symmetry and, therefore, would not be expected to demonstrate optical activity.

Another important example of the significance of demonstrating optical activity is the work of Mills and Quibell (2), who showed that the platinum(II) ion is not tetrahedral by resolving the *racemic* (*meso* - stilbenediamine)(*iso* - butylenediamine)platinum(II) cation, DL-[Pt(*m*-stien)(*i*-bn)]²⁺ (2). This work gave support for the proposed

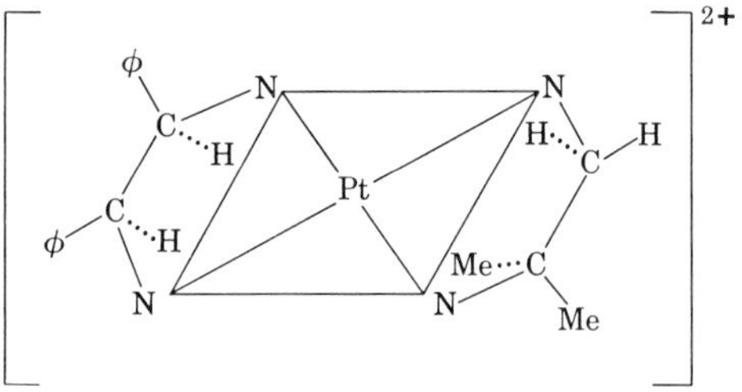

2

planar form of the ion; this form would not contain a plane of symmetry and, would, therefore, be expected to be optically active, whereas the tetrahedral configuration (**3**) would contain a plane of symmetry and

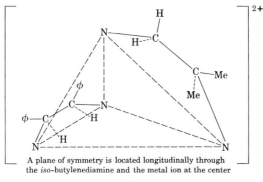

A plane of symmetry is located longitudinally through
the *iso*-butylenediamine and the metal ion at the center

3

therefore, would not exhibit optical activity.

A third example of the importance of being able to resolve racemic modifications is found in the work of Dhar, Doron, and Kirschner (3,4), who demonstrated that silicon(IV) can exist as a 6-coordinate species by resolving the *racemic* tris(acetylacetonato)silicon(IV) cation, DL-[Si-(acac)$_3$]$^+$, into its optical enantiomers, thereby confirming that the structure of this complex ion is similar to that of (**1**). All the other reasonable structures (e.g., tetrahedral, planar, hexagonal, prismatic, etc.) which might be proposed for this complex would possess a plane of symmetry and would, therefore, not be expected to exhibit optical activity.

Other illustrations of the importance of developing techniques for the resolution of racemic modifications of complexes are to be found in the areas of optical rotatory dispersion (5–9) and reaction mechanisms (10,11).

II. OPTICAL ACTIVITY, POLARIMETRY, AND SPECTROPOLARIMETRY

A. Optical Activity

The phenomenon of optical activity was discussed as early as the beginning of the nineteenth century when Arago (12) studied the behavior, in quartz and other minerals, of sunlight which was polarized by reflection. Biot (13) studied this phenomenon in some detail and noted that the plane of polarization of linearly polarized light was rotated by quartz crystals and by solutions of certain naturally occurring substances such as camphor, sugar, etc. Later developments led to the correlation of the ability of a substance with an asymmetric configuration to rotate the plane of polarized light and, later, finer distinctions were made to indicate such correlations with "dissymmetric" configurations. Strictly speaking, a molecule need not be totally without symmetry (asymmetric) in order to demonstrate optical activity. Molecules which lack a center of symmetry and a plane of symmetry are usually optically active, even though they exhibit certain elements of symmetry (e.g., an axis of symmetry). Such molecules are called "dissymmetric" and this term is to be preferred when referring to most optically active complexes, except, of course, any which are totally devoid of all symmetry.

The theoretical bases of optical activity have undergone much study since its discovery and are discussed in several treatises (14–17). It is important to note that the demonstration of optical activity in a compound is used as confirmatory evidence that the compound lacks a plane of symmetry. This fact has important stereochemical and structural connotations and there is considerable interest in utilizing optical rotation, when feasible, as a tool in structural studies.

B. Polarimetry

Because of a resurgence of interest in optical rotation, in general, and in optical rotatory dispersion (the wavelength dependence of optical rotation), in particular, a large number of devices (polarimeters) for measuring the rotation of plane polarized light by substances has become available. There are wide variations in type (visual, photoelectric, photographic, etc.), basis of operation (direct reading, Faraday effect, absorbance-based, etc.), flexibility (single-wavelength, multi-wavelength spectropolarimeters, etc.) precision, and cost. Several articles discuss many of these instruments and describe their operation, advantages, and disadvantages in some detail (8,14,17a).

Most visual determinations of optical rotation at a single wavelength are still carried out at the sodium "D" line (589 mμ), although the mercury green line (546 mμ), is also used quite often and is frequently preferred because of the greater sensitivity of the eye to green over yellow. The sample is usually a solution which is placed in a 1 or 2 dm. polarimeter tube at a temperature of either 20 or 25°C. Optical rotation is almost always expressed by chemists as "specific rotation" ($[\alpha]$) or "molecular rotation" ($[\alpha]_M$) in degrees according to the following:

$$[\alpha] = \alpha/lc$$
$$[\alpha]_M = M[\alpha]/100$$

(with the temperature indicated to the upper right and the wavelength to the lower right of the bracket), where l is the length of the sample (dm.); c is the concentration of the solution sample in g./ml. of solution; M is the molecular weight of the sample; and α is the observed rotation in degrees.

There is some disagreement in the literature regarding the definition of molecular rotation ($[\alpha]_M$) since some investigators do not divide by 100 as shown in the equation (18,19), but most research workers in polarimetry accept the definitions given above (14).

In practice, an optical rotation reported at a single wavelength which has been determined on a visual instrument is the average of an odd number of individual determinations (usually at least five) and is reported with the average deviation. An operator working with a high precision visual instrument under good conditions can often achieve an average deviation of $\pm 0.003°$. Electronic instruments have been perfected which surpass this precision by an order of magnitude, however. For compounds which absorb light to a considerable extent at the sodium or mercury lines, it is recommended that samples be diluted, or, if this is not feasible because of low rotations, that light sources be combined with filters or monochromators allowing the use of different wavelengths.

C. Spectropolarimetry

The recent increase in interest in optical rotatory dispersion has led to the development of several types of instruments for its determination (8,17). Basically an optical rotatory dispersion curve is a plot of optical rotation vs. wavelength for a given substance (Fig. 1).

If an optical rotatory dispersion curve exhibits the behavior shown in Figure 1 in the vicinity of an absorption band, the absorption band is referred to as an "optically active absorption band" and the behavior is referred to as "anomalous rotatory dispersion" and is sometimes called the "Cotton Effect" (14). It should be emphasized that at certain wavelengths a single enantiomer of an optically active complex may have its optical rotatory dispersion pass through zero degrees and therefore will show no optical rotation at those wavelengths.

In comparing the optical rotations of different coordination compounds, it is sometimes useful to employ the "equivalent molecular rotation "($[\alpha]_e$ or $[M]_e$ or $[E]$) in order to avoid nonmeaningful differences in molecular rotation due to differences in molecular weight of the optically *inactive* ion associated with the optically active complex ion. These differences might occur when making comparisons of optical

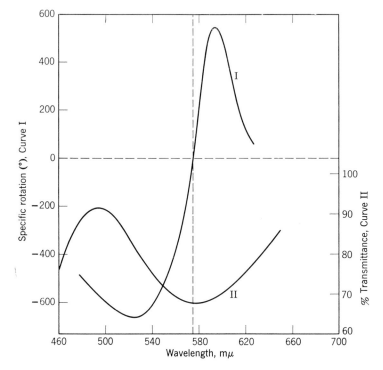

Fig. 1. I:Optical rotatory dispersion of D-K$_3$[Cr(C$_2$O$_4$)$_3$]; II:optical
absorption spectrum of D-K$_3$[Cr(C$_2$O$_4$)$_3$].

rotations (e.g., comparing the optical rotation of [Co(en)$_3$]$_2$(SO$_4$)$_3$ with
that of [Co(en)$_3$]Cl$_3$). Equivalent molecular rotation is defined as (8):

$$[M]_e = M[\alpha]/100n = [\alpha]_M/n$$

where n is the number of optically active complex ions per formula
unit of the coordination compound.

III. TECHNIQUES FOR OBTAINING OPTICALLY ACTIVE COORDINATION COMPOUNDS

There are several general methods by which optically active coordination compounds may be prepared. The methods most often used
go under the heading of "Resolution Procedures," and involve separation
of enantiomers from a racemic mixture of isomers of opposite (mirror-
image) configurations. However, it is also possible to obtain optically
active complexes by direct methods which do not involve separation or

resolution of a mixture containing the desired enantiomer. Strictly speaking, resolution procedures are those which are capable of separating both enantiomers of a racemic mixture with ultimate recovery of both isomers, although many authors who have been able to recover only one enantiomer by a given procedure often refer to the work as a "resolution." In this discussion the term *resolution techniques* will refer to those techniques for which, in principle, both enantiomers of a racemic mixture are at least partially obtainable.

A. Resolution Procedures

Since the time of Pasteur's early work in the field of optical activity, a large number of techniques have been devised for the resolution of racemic modifications and most of these have been applied to complex inorganic compounds. Many of the methods are designed for the complete separation of enantiomers, but several techniques result in partial separations only—usually just sufficient separation to establish optical activity and, therefore, asymmetry (or dissymmetry) in the complexes in question.

1. Mechanical Separation of Antipodes

The earliest method of resolving racemic mixtures was developed by Pasteur (20) in 1848 and involved the handpicking of crystals exhibiting different hemihedral faces, usually with the aid of a microscope or magnifying glass. Pasteur did the original work with a racemic mixture of the enantiomers of sodium ammonium tartrate. The method requires facility in crystal growing and in the identification of hemihedral faces and is successful only with racemic *mixtures* as opposed to racemic *compounds*, twinned crystals, etc. Because it is usually very time consuming, requires exacting manipulations, and produces satisfactory results only in certain circumstances, the method is not widely used today as a technique for resolving racemic modifications.

2. Spontaneous Crystallization of Racemic Mixtures

In 1919, Jaeger (21) demonstrated a technique for the conversion of a racemic compound into a racemic mixture by its crystallization above a particular temperature which is dependent upon the complex being studied. An example, which is a variation of Pasteur's method, is the crystallization of $K_3[Co(C_2O_4)_3]$ from aqueous solution above 13.2°C. giving a mixture of the *dextro* and *levo* enantiomers that can be resolved by mechanical separation.   An equation to represent the conversion is*

$$2K_3[Co(C_2O_4)]\cdot3.5H_2O \underset{13.2\,^\circ C.}{\overset{}{\rightleftharpoons}} \text{D-}K_3[Co(C_2O_4)_3]\cdot H_2O + \text{L-}K_3[Co(C_2O_4)_3]\cdot H_2O + 5H_2O$$

Jaeger (6) also obtained a racemic mixture of a cyclopentanediamine complex, [Rh(cptn)$_3$](ClO$_4$)$_3$·12H$_2$O, by spontaneous crystallization techniques below 48 °C. and was also able to resolve these isomers by mechanical means.   In addition to suffering the disadvantages mentioned in Sec. 1, the method also requires the determination of the proper temperature for crystallization of a mixture of enantiomers rather than the racemoid.   Consequently, this method is also little used today.

3. Preferential Crystallization by Isomorph Seeding

The usual object in the resolution of racemic mixtures of complexes is to separate the enantiomers.   An interesting approach to this problem was studied by Werner and Bosshart (22a) with racemic mixtures of the [Co(en)$_2$(C$_2$O$_4$)]$^+$, [Cr(en)$_2$(C$_2$O$_4$)]$^+$, and [Co(en)$_2$(NO$_2$)$_2$]$^+$ ions. They found that if a seed crystal of the D-[Co(en)$_2$(C$_2$O$_4$)]$^+$ ion was added to solutions of racemic mixtures of these ions, one enantiomorph crystallized preferentially.   After addition of the seed crystal, it was usually necessary to add a mixture of ethanol and ether in order to effect the desired crystallization.   Further, the filtrate was considerably enriched in the opposite enantiomorph.

Because some of the techniques described later have met with considerable success, relatively little work has been done on this technique, although it appears to offer considerable potential as a resolution method, since apparently only a seed crystal of an isomorph of the desired enantiomer is required and not necessarily a crystal of the desired isomer itself (22b).

The method also has potentialities for the determination of absolute configurations of optically active complex ions, and should certainly provide at least supporting evidence for absolute configurations of complex ions which are successfully crystallized by seed crystals of known configuration.

4. The Formation of "Active Racemates"

Delépine (23,24a) developed a resolution technique which is based upon the fact that racemic modifications in the solid state may exist in three possible forms that are sometimes substantially different from

* Upper case D and L refer to the signs of rotation for complex ions at the wavelength studied, and lower case d and l refer to the signs of rotation of ligands, with no reference being made to the absolute configurations of these substances unless otherwise stated.

the enantiomers from which they are formed. The three forms are known as *racemic mixtures, racemic compounds,* and *racemic solid solutions.* Racemic mixtures result from asymmetric substances which form in crystals containing hemihedral facets which are enantiomorphic. It is this type of mixture which is susceptible to resolution by mechanical picking-out of crystals containing only one enantiomer. A *racemic compound,* however, is produced when a pair of enantiomers combine to form a substance in which all the crystals are identical and each crystal contains equal quantities of each enantiomer. Such crystals have properties different from those of the crystals of the individual enantiomorphs. A *racemic solid solution* may crystallize without compound formation when two enantiomorphs happen also to be isomorphous.

Delépine recognized that similar asymmetric (or dissymmetric) complex ions might have the same relative configuration even though their optical rotations at a given wavelength might be entirely different. A combination of one such ion with the "opposite" enantiomer of the other might therefore be expected to form a racemic compound or racemic solid solution instead of a racemic mixture, which is the basis of the method of "active racemates." For example, when a solution containing equimolar quantities of D- and L-$K_3[Co(C_2O_4)_3]$ is allowed to evaporate, a *racemic mixture* crystallizes. If the crystals are dissolved in water, the resulting solution is optically inactive. However, if a solution containing equimolar quantities of D-$K_3[Cr(C_2O_4)_3]$ and L-$K_3[Co(C_2O_4)_3]$ is allowed to evaporate, crystals are formed which yield an optically active solution (the *"active racemate"*) because the chromium complex does not have an optical rotation of exactly the same magnitude as that of the cobalt complex. Should the active racemate be a *racemic mixture,* then the relative configurations of the enantiomorphs are different, but if the active racemate proves to be a *racemic compound* or *racemic solid solution,* then the relative configurations of the components are the same and a basis for a resolution procedure exists for such cases.

Delépine was, in fact, able to demonstrate that active racemates which are racemic compounds or racemic solid solutions actually form between the following pairs: L-$K_3[Ir(C_2O_4)_3]$ and D-$K_3[Rh(C_2O_4)_3]$, D-$K_3[Ir(C_2O_4)_3]$ and L-$K_3[Co(C_2O_4)_3]$, and D-$K_3[Ir(C_2O_4)_3]$ and L-$K_3[Cr(C_2O_4)_3]$. Therefore, taking the last pair as an example, a technique for resolving D,L-$K_3[Cr(C_2O_4)_3]$ would consist of treating a solution of the racemic mixture with D-$K_3[Ir(C_2O_4)_3]$, evaporating to allow crystallization of the racemic compound (the "active racemate"), and recovering the D-$K_3[Cr(C_2O_4)_3]$ from the supernate, according to the following:

D-$K_3[Ir(C_3O_4)_3]$ + D,L-$K_3[Cr(C_2O_4)_3]$ =
$$\text{D-}K_3[Ir(C_2O_4)_3]\cdot\text{L-}K_3[Cr(C_2O_4)_3] + \text{D-}K_3[Cr(C_2O_4)_3]$$
"active racemate"

Subsequent separation of the active racemate by other techniques would yield the other enantiomer and would complete the desired resolution.

5. Configurational Activity Effect Techniques (First-Order Asymmetric [or Dissymmetric] Transformation) (24b)

Dwyer and his co-workers (25,26) utilized the fact that while optically inactive electrolytes added to a solution of a racemic mixture of an optically active complex compound altered the activity of each enantiomer identically, the addition of an optically active ion had a slightly different effect on the activity of each of the enantiomers. Dwyer observed solubility differences up to the extent of 3.5% for D- and L-[Ru(o-phen)$_3$](ClO$_4$)$_2$ (o-phen = ortho-phenanthroline) in dilute solutions containing ammonium d-bromo-π-camphor sulfonate or Rochelle salt (sodium potassium d-tartrate). He observed similar effects for racemic mixtures of the optically active complexes [Ni(dipy)$_3$]I$_2$ (dipy = 2,2'-dipyridyl) and [Co(acac)$_3$] (acac = acetylacetonate anion). He attributed this configurational activity effect to the different interactions of the electronic fields of the different enantiomers with the field of the optically active environment created by addition of the optically active substance. At higher concentrations the effect diminishes, probably because the normal activity effects tend to overwhelm the relatively slight configurational activity effects at high ionic strengths. This effect has also been observed in other systems (11).

6. Biological Methods

This technique was also developed by Pasteur (27) who observed that when either yeast or the mold *Penicillium glaucum* was allowed to ferment the ammonium salt of racemic tartaric acid, the salt of the naturally occurring acid (the *dextro* isomer) was consumed preferentially and, ultimately, the salt of the *levo* isomer could be recovered in an optically pure form from the fermentation broth. Enzymes will also catalyze similar resolutions. For example, hog-kidney acylase (*"acylase I"*), when acting on an acylated racemic amino acid in aqueous solution until half the acyl groups are hydrolyzed, will produce a residual amino acid which is the derivative of the unnatural isomer (d) and the free amino acid which is the natural (l) enantiomer (28,29). In principle, there should be available enzymes and bacteria which could act on each enantiomer of a racemic mixture, thus, theoretically allowing at least partial resolution.

With regard to coordination compounds, Crespi (30) attempted a partial resolution of D,L-[Co(en)$_3$]$^{3+}$ by allowing a particular organism to feed upon one enantiomer until it "learned" to assimilate that isomer and then allowing these bacteria to feed upon the racemic mixture of the complex. The technique has not 'yet been proved to be successful for resolving complexes, although it should be applicable to other resolutions, particularly with complexes which exhibit marked biological activity.

7. The Formation and Fractional Crystallization of Diastereoisomers (Second-Order Asymmetric [or Dissymmetric] Transformations)

This is by far the most important and most frequently used technique for the resolution of racemic substances. Basically the method consists of treating a racemic mixture of an ionic complex with a single enantiomer of an optically active "resolving agent" of opposite charge with the resultant formation of nonmirror image diastereoisomers. For example,

$$\text{D-}[Co(en)_3]Cl_3 + \text{L-}[Co(en)_3]Cl_3 + 2Na_2d\text{-tart} =$$
$$\text{D-}[Co(en)_3]Cld\text{-tart} + \text{L-}[Co(en)_3]Cld\text{-tart} + 4NaCl$$

The most common technique for separating these diastereoisomers is by fractional crystallization because crystal structure is a property which is particularly sensitive to minor variations in molecular structure, such as occur in diastereoisomers. Other techniques have been employed to effect the separation of diastereoisomers including distillation (31), chromatography, zone melting, etc., but fractional crystallization remains the most important. Some of the other techniques will be discussed later (vide infra).

The optically active ions most frequently used as resolving agents for racemic mixtures of complex ions are listed in Table I. Also listed are some of the adsorbents which have been utilized in the resolution of optically active inner (non-ionic) as well as ionic complexes.

This method of diastereoisomer formation is not directly applicable to the resolution of racemic mixtures of inner complexes unless they possess a functional group which is capable of interacting with a resolving agent. An example of a resolution of such a complex is found in the use of strychnine for the resolution of racemic-bis(8-quinolino-5-sulfonic acid)zincate(II) (36).

In order to effect complete resolution, it is necessary to provide for removal of the resolving ion once the diastereoisomers have been separated, so as to generate the original enantiomers as separate substances. Various techniques have been developed for this procedure. For example, when alkaloids are used as resolving agents, the complex diastereoiso-

TABLE I

Some Resolving Agents Used for the Resolution of Racemic Complex
Inorganic Compounds

Cations, for complex anions	Anions, for complex cations	Adsorbents, for non-ionic and ionic complexes
Strychnine (as the hydro cation)	Tartrate	Quartz (optically active)
Brucine (as the hydro cation)	Antimonyl tartrate	$NaClO_3$ (optically active)
Cinchonine (as the hydro cation)	α-Bromocamphor-π-sulfonate	Starch
α-Phenylethylamine (as the hydro cation)	Dibenzoyl-d-tartrate (32,33)	Coated alumina
Cinchonidine (as the hydro cation)	Camphor-π-sulfonate	d-Lactose hydrate
Quinine (as the hydro cation)	Nitrocamphoronate	Cellulose
Quinidine (as the hydro cation)	Diacetyltartrate	Glucose
Morphine (as the hydro cation)	$[Co(edta)]^-$ (34)	Sucrose
cis-$[Co(en)_2(NO_2)_2]^+$ (35)		Tartaric acid
		Sodium glutamate
$[Co(en)_3]^{3+}$		Paper
$[Ni(o\text{-phen})_3]^{2+}$		Coated silica gel

mers are frequently ground with solid potassium iodide or sodium perchlorate and then the resultant mixture is treated with water to extract the
alkali metal salt of the resolved complex anion, leaving behind the relatively insoluble alkaloid iodide or perchlorate. Occasionally the same
diastereoisomer can be treated with base to generate the free alkaloid,
which is usually relatively insoluble in water. Resolving cations are
frequently removed by precipitation techniques or by treatment with
mineral acid to generate the resolving agent in the often relatively insoluble form of its free acid. Newer techniques involve the utilization
of ion exchange resins for removal of resolving ions (3,4), but caution
must be applied to insure complete removal of the resolving ion by the
resin.

8. Displacement Techniques (Kinetic Asymmetric [or Dissymmetric] Transformation)

In 1952, Bailar and his co-workers (37,38) observed a configurational
influence on the kinetics of the displacement of carbonate ion from
$[Co(l\text{-pn})_2CO_3]Cl$ (l-pn = levo-propylenediamine) by racemic tartaric,
α-chloropropionic, and lactic acids; and, in 1955, Dwyer and his co-
workers (39) reported that ethylenediamine displaced the ethylene-

diaminetetraacetate ion from D-[Co(edta)]⁻ with retention of optical activity. Kirschner and his co-workers (40) utilized this information to effect a partial resolution of D,L-[Co(edta)]⁻ by preferential displacement of the ethylenediaminetetraacetate ion from one enantiomer with *levo*-propylenediamine, according to the reaction:

$$\text{D-[Co(edta)]}^- + \text{L-[Co(edta)]}^- + 3\ l\text{-pn}$$
$$= \text{L-[Co(}l\text{-pn)}_3]^{3+} + \text{D-[Co(edta)]}^- + \text{edta}^{4-}$$

Das Sarma and Bailar (41), after treating D,L-*cis*-[Co(trien)Cl₂]Cl (trien = triethylenetetramine) with silver *d*-antimonyl tartrate, obtained L-[Co(trien)(*d*-SbOtart)₂](*d*-SbOtart), which, upon conversion to the chloride, gave L-*cis*-[Co(trien)Cl₂]Cl, although the degree of its optical purity was not determined.

This technique is of considerable interest and is being studied by other investigators as a potential method for the resolution of other complexes.

9. Photochemical Techniques

Since enantiomers of opposite configuration absorb circularly polarized light of one handedness (i.e., either right or left circularly polarized light) to different extents, it would be expected that the two enantiomers of a racemic mixture should undergo photodecomposition at different rates upon exposure to such circularly polarized light of an appropriate wavelength. Kuhn and co-workers (42,43) obtained positive results with such techniques for organic compounds, but an attempt by Jaeger and Berger with the D,L-[Co(C₂O₄)₃]³⁻ ion gave negative results (44a). Tsuchida and co-workers (44b), however, report that they obtained a partial resolution of D,L-K₃[Co(C₂O₄)₃] in aqueous solution by the preferential destruction of the dextrorotatory enantiomer by circularly polarized light. This technique deserves further study as a potential means of resolving complexes, especially in light of the recent developments in instrumentation which should facilitate experimentation in this area.

10. Preferential Adsorption Techniques

a. Powdered Optically Active Quartz. Investigators interested in the resolution of racemic mixtures of non-ionic inner complexes have developed an important and interesting technique which is based upon the ability of certain substances to adsorb selectively (i.e., either to a greater extent or at a greater rate) one enantiomer of the mixture. The method was initially used by Tsuchida and co-workers (45) to achieve a partial resolution of D,L-[Co(dmg)₂(NH₃)Cl] (dmg = dimethylglyoximino anion) by shaking a solution of the substance with powdered *levo*-

quartz. Several other investigators (46) have utilized this technique with an isomer of quartz with varying degrees of success. Although no instances of complete resolution by this technique have yet been confirmed (however, see ref. 46d), it has been extremely useful in demonstrating optical activity in several complex compounds, both ionic and nonionic. It has the advantage of avoiding chemical operations on an optically active system (e.g., diastereoisomer formation) and of permitting rapid transfer of an optically active enantiomer to a polarimeter for rapid determination of activity in cases of relatively labile substances.

   b. Powdered Optically Active Sodium Chlorate. In addition to quartz, several other optically active adsorbents have been used in an effort to effect resolutions of racemic mixtures (*vide infra*). With few exceptions, these adsorbents are optically active because of the presence of one or more asymmetric carbon atoms and not because they possess molecular asymmetry (or dissymmetry) without asymmetric carbon atoms, such as occurs in optically active quartz. One exception is the report by Ferroni and Cini (47) in which a partial resolution of D,L-$[Be(bzac)_2]$ (bzac = benzoylacetonate anion) is described by preferential adsorption on both *dextro* and *levo* powdered sodium chlorate, in which the optically active large single crystals were grown over several days from aqueous super-saturated solutions of the salt. Sodium chlorate is optically active because it possesses, as does quartz, an unsymmetrical molecular configuration in the crystal. There are several other substances which possess naturally optically active lattices in the crystalline state. Several of these are under study as possible optically active adsorbents for the resolution of racemic mixtures of complexes by preferential adsorption.

   c. Powdered Starch. Krebs and his co-workers (48) have studied starch as an adsorbent for the preferential adsorption of enantiomers and have achieved partial resolutions of several racemic mixtures by this technique, including D,L-$K_3[Cr(C_2O_4)_3]$ and D,L-α-$[Co(glycinate)_3]$.

   d. Coated Alumina. Karagounis and co-workers (49) prepared an adsorbent of aluminum(III) oxide coated with various optically active substances in attempts to resolve racemic mixtures of optically active compounds. Several of these attempts were successful, and Krebs and Rasche (48a) and Piper (50) also applied the technique to the resolution of racemic mixtures of complex compounds.

   e. d-Lactose Hydrate. This adsorbent has been used by Moeller and co-workers (51) and Collman and co-workers (10) for the resolution of several inner complexes, such as the tris(acetylacetonates) of trivalent metal ions.

f. Vapor-Phase Techniques. Karagounis and Lippold (63) and Sievers and co-workers (64) have utilized vapor-phase chromatographic techniques to effect resolutions of racemic mixtures of optically active organic and complex inorganic compounds, respectively. Sievers utilized a column containing powdered *d*-quartz and resolved D,L-[Cr(hexafluoroacetylacetonate)$_3$] by gas–solid chromatography.

g. Other Adsorbants. Several other substances have been utilized as adsorbents in attempts to resolve racemic mixtures of optically active compounds. Not all have been successful, nor have all been used with complexes. Cellulose has been used by Krebs and co-workers (48d) and by Mayer and Merger (52); ion exchange resins with optically active groups have been used by Krebs and co-workers (48d) and others (53,54); Karagounis and co-workers (49) have used *d*-fenchone, *d*-glucose, *d*-tartaric acid, sodium-*l*-glutamate, and *d*-lactose; paper chromatography has been utilized by Buchar and Suchy (55) for complexes and by others (56,57) for organic compounds; silica gel coated with optically active substances has been used by Curti and Colombo (58) and by Klemm and Reed (59); and sucrose has been used by Carassiti (60). The technique has also been used by Jamison and Turner and others (61). Other techniques related to these are under study and should yield fruitful results. Among them are the preferential inclusion of a single enantiomer in an optically active "cage," such as might be considered to form when urea crystallizes, as well as the destruction of single enantiomers by catalysts containing optically active substrates such as *d*-quartz.

11. Zone-Melting Techniques

Doron and Kirschner (62) applied zone-melting techniques to the partial resolution both of mixtures of diastereoisomers and of racemic complexes at low temperatures. For the former, the solvent used was water and for the latter, a mixture of water and dioxane containing an optically active salt which did not react directly with the enantiomers, thereby utilizing the configurational activity effect of Dwyer for the resolution. The apparatus was housed almost entirely in a deep freeze unit and had the advantage of requiring no attention during most of the operation. The method is based upon the principle that a mixture of diastereoisomers (or of enantiomers in an optically active environment) in a column of frozen solution will dissolve at unequal rates in a molten zone which traverses the column. The molten zone, therefore, becomes enriched in the substance which dissolves at the more rapid rate, resulting in enrichment of this substance at one end of the column of frozen solvent and relative enrichment of the other substance at the other end.

B. Nonresolution Techniques

1. Equilibrium Methods of Enantiomer Production

a. Solubility Differences. In attempting resolutions of certain racemic mixtures, it was found that precipitation techniques sometimes yielded practically 100% of one enantiomer, even though the original mixture contained only 50% of that isomer and 50% of its mirror image form. The result occurs because of the existence of a dynamic equilibrium between the enantiomers in solution, and precipitation of one enantiomer as a diastereoisomer causes a rapid displacement of that equilibrium in favor of the lost enantiomer, which is then further precipitated by addition of more resolving agent until complete conversion to the enantiomer forming the less soluble salt with the resolving agent is effected.

Werner (65), in attempting to resolve D,L-$K_3[Cr(C_2O_4)_3]$ with strychnine, noted that in ethanol solution only the dextro-rotatory enantiomer was obtained, whereas in water only the levo-rotatory isomer was obtained, undoubtedly because of the different solubilities of the strychnine salts of these enantiomers in the two solvents. Dwyer and Gyarfas (66) noted that a solution of D,L-$Fe(o$-phen$)_3]^{++}$ containing excess *d*-KSbOtart produced a precipitate in which *all* of the complex iron compound was present as L-$[Fe(o$-phen$)_3]d$-$SbOC_4H_4O_6\cdot4H_2O$, because of the effect of the dynamic equilibrium coupled with the much lower solubility in water of this diastereoisomer than the one containing the *dextro* enantiomer of the iron complex. Other examples of this type of enantiomer production are also in the literature (67).

b. Oxidation–Reduction Procedures. Another related technique by which it is possible to bring about enantiomer formation in "yields" greater than expected has been described by Busch (68) who treated D,L-$[Co(en)_3]^{3+}$ with excess *d*-tartrate, thereby precipitating a fraction of the dextrorotatory isomer of the complex. Racemization of the filtrate was then achieved by electron transfer equilibration with $[Co(en)_3]^{++}$, producing additional dextro-rotatory isomer which was then precipitated by the *d*-tartrate. Ultimately, three-fourths of the cobalt complex was obtained in the dextro-rotatory form.

Dwyer and Gyarfas (66,69a) also developed procedures for the preparation of optical enantiomers by redox techniques. For example, D- and L-$[Ru(o$-phen$)_3](ClO_4)_2$ were independently oxidized with cerium-(IV) nitrate to the tripositive oxidation state of ruthenium with retention of optical activity, so all four enantiomers were obtained in a high degree of optical purity. Further, they found it possible to reduce the tripositive enantiomers with iron(II) sulfate and they obtained the dipositive species again with retention of optical activity (70). Im and Busch (71) showed that optically active propylenediaminetetraacetate complexes of cobalt(II) and cobalt(III) could be formed by oxidation reactions analogous to those described above.

2. Asymmetric Synthesis and Dissymmetric Synthesis

a. *Absolute (Total) Asymmetric Synthesis.* General agreement does not exist with regard to the definitions of the terms *asymmetric synthesis* (*dissymmetric synthesis*, if the optically active compound formed contains elements of symmetry), *absolute asymmetric synthesis*, and *partial asymmetric synthesis* (72). Fischer (73) and Marckwald (74) regard asymmetric synthesis as a process by which optically active substances are synthesized from optically inactive species by means of optically active reagents as intermediates only, and without the use of resolution procedures. Absolute asymmetric synthesis is generally regarded as a process whereby an optically active molecule is produced from optically inactive substances without the use of either resolution procedures or optically active *reagents*, but this does not exclude the use of such things as circularly polarized light, etc. Although an absolute asymmetric synthesis has been recognized for organic compounds (75), as has an *asymmetric destruction* by circularly polarized light (43,44b), the absolute asymmetric synthesis of a coordination compound has yet to be achieved.

b. *Partial Asymmetric Synthesis and Partial Dissymmetric Synthesis.* A partial asymmetric synthesis (partial dissymmetric synthesis, for optically active compounds containing elements of symmetry) is generally regarded as the synthesis of an optically active molecule from an optically inactive one without the use of a resolution procedure, but with the use of an optically active reagent which becomes a part of the product molecule. The fundamental feature is that the optically active reagent induces a preferred optically active configuration in the product molecule of which it becomes a part, for example, the platinum (IV) complex containing propylenediamine, $[Pt(l\text{-}pn)_3]^{4+}$. It appears that the configuration (and rotation) of the cation as a whole can be influenced by the presence of optically active ligands and that frequently only one isomer of a complex is formed (76,77) when only one enantiomer of a ligand is used.

Another example can be found in the reaction between cobalt(II) and *d-trans*-1,2-diaminocyclopentane which (upon oxidation) yields practically exclusively the enantiomer L-$[Co(d\text{-cptn})_3]^{3+}$ (78).

3. *Displacement Techniques*

Certain displacement reactions on optically active enantiomers of complexes occur with retention (or inversion) of optical activity and provide a method for the synthesis of optical isomers. Dwyer and co-workers (39) reported the formation of D-$[Co(en)_3]^{3+}$ upon treatment of D-$[Co(edta)]^-$ with ethylenediamine; Kirschner and co-workers (40) reported the formation L-$[Co(l\text{-pn})_3]^{3+}$ from the reaction of D-$[Co(edta)]^-$ with *levo*-propylenediamine; and Messing and Basolo (79) oxidized L-$[Pt(m\text{-stien})(i\text{-bn})]^{++}$ (*m*-stien = *meso*-stilbenediamine; *i*-bn = *iso*-butylenediamine) and obtained L-$[Pt(m\text{-stien})(i\text{-bn})Cl_2]Cl_2$, which, on reduction, gave the original platinum(II) complex with its original optical rotation.

An interesting example of this technique was presented by Bailar and Auten (80) who observed that treatment of L-$[Co(en)_2Cl_2]^+$ with potassium carbonate gave D-$[Co(en)_2CO_3]^+$ and with silver carbonate gave either D- or L-$[Co(en)_2CO_3]^+$, depending upon the reaction conditions, which indicated that a form of Walden inversion had occurred in certain of these reactions.

IV. OPTICAL PURITY

A. Definitions

The optical purity of a partially resolved material is defined as the excess of one enantiomer in the material expressed as a percentage of the total. For a case where a *dextro* enantiomer is present in excess:

$$O.P. = (E/T)(100) = [(D - L)/(D + L)](100)$$

where $O.P.$ is the optical purity (in per cent), E is the excess of one enantiomer, T is the total quantity of optically active material, D and L are quantities of *dextro* and *levo* enantiomers, respectively. In a D,L pair there is no excess of one enantiomer and the optical purity is zero. For a completely resolved substance the "excess" enantiomer is equal in weight to the total, so the optical purity is 100%.

B. Criteria for Optical Purity

The criteria which are most commonly used as an indication of the complete optical purity of a crystalline substance are that its melting point and optical rotation are unchanged after further crystallization.

However, these criteria are not to be taken as infallible, since certain solid solutions may give partially resolved mixtures which do not change melting point or rotation upon recrystallization. Another criterion which is often used is the formation of enantiomers having equal specific rotations of opposite sign, but such a criterion should be regarded with caution unless the same enantiomers possessing the same specific rotations are obtained by two or more different techniques.

C. Determination of Optical Purity

There are three primary techniques for the determination of the optical purity of optically active enantiomers. Although none of these has been utilized for the specific purpose of optical purity determination for complex inorganic compounds, they are applicable in principle to this purpose.

1. Biological Technique

If an enzyme, bacterium, or other organism is selective for one enantiomer of a racemic pair, then incubation of the opposite isomer with the biological substance should produce no reaction. The occurrence of reaction would indicate an impure isomer. A suitable method of detection of reaction would, of course, have to be developed for each system.

2. Correlative Reaction Technique

This technique provides for the determination of the *minimum* degree of optical purity of an enantiomer preparation, and requires that there be available a resolved substance of known optical purity, e.g., D-[M(abcd)]. If it is desired to obtain the optical purity of, for example, D-[M(abce)], which can be converted chemically to D-[M(abcd)], the degree of optical purity of the former will be *at least* as great as that of the latter from which the former was prepared. D-[M(abce)] may have a *greater* degree of optical purity than D-[M(acbd)] because some racemization may occur during the chemical reaction (81).

3. Isotope Dilution Technique

This technique consists of mixing in solution a supposedly pure enantiomer (say, a D isomer) with some radioactive or otherwise labeled racemic material, re-isolating the racemic material, and then comparing the calculated (e.g., radioactive) dilution factor with that observed.

The observance of a lower dilution factor than that calculated assuming initially pure D enantiomer indicates that some of the opposite enantiomer had been present, and the amount can actually be calculated (82,83). If the D enantiomer were pure, mixing it with labeled racemic material would result, upon re-isolation of the racemic material, in a dilution of radioactivity of the D isomer only, but not of the L. It is thus possible to calculate the expected dilution factor from the weights of the original enantiomer and of the added labeled racemic material. If the original material contained *both* unlabeled D and L enantiomers, then mixing this with labeled racemic material results in a dilution of *both* enantiomers and a greater dilution factor than before. Mathematically,

$$C_{\pm} = [aC_0](a + B)/(2B + a - R)(a + R)]$$

where $C_{\pm}$ is the activity of the re-isolated racemic mixture, C_0 is the activity of the added racemic material, a is the weight of the added racemic material, B is the weight of the resolved enantiomer (whose optical purity is in question) and which is mixed with a, and R is the weight of a racemate (if any) in B. Solving the above equation for R will allow calculation of the optical purity (in per cent) according to

$$O.P. = [(B - R)/B](100)$$

V. EXPERIMENTAL

Most of the literature cited in the preceding sections contains experimental data and instructions for carrying out the procedures described. In addition, a later section of this chapter will contain a listing of several optically active complex inorganic compounds which are obtainable in enantiomeric form along with the appropriate literature citations which give experimental details. Further, a suggested bibliography will be presented which will enable the reader to locate sources of experimental details for the preparation of many enantiomers of coordination compounds.

With regard to several of the techniques discussed, specific mention can be made of certain experimental procedures for the techniques in most common use. Concerning the technique of preferential adsorption, a recent paper (10) gives explicit instructions for the construction of a reusable adsorption column filled with *d*-lactose hydrate. With regard to the technique of configurational activity, the papers by Dwyer and co-workers (25,26) may be consulted for experimental details.

For details regarding equilibrium techniques of enantiomer formation, several papers (66,68,69,71) can be cited, which is also the case for displacement techniques (39,40,79,80).

The most frequently used technique involves diastereoisomer formation and the many references cited should be consulted for experimental details. Several excellent resolution procedures utilizing this technique are reported in the various volumes of *Inorganic Syntheses*. An interesting example of one such resolution is a modification of the procedure of Dwyer and Garvan (35) for the resolution of the *racemic-cis*-dinitrobis(ethylenediamine)cobalt(III) ion, which is carried out as follows:

Resolution of D,L-*cis*-$[Co(en)_2(NO_2)_2]^+$.   Dissolve 12.0 g. of D,L-*cis*-$[Co(NO_2)_2]_2]NO_2$ (84a) in 50 ml. of water at 60°C. with vigorous stirring in a 250 ml. Erlenmeyer flask.   Quickly add a solution composed of 7.0 g. of potassium antimonyl-*d*-tartrate (d-$KSbOC_4H_4O_6$) dissolved in 40 ml. of water at 75°C. by vigorous shaking.   Cool the mixture quickly to 25°C. with tap water and scratch the inside of the flask under the surface of the liquid with the end of a glass stirring rod which is not fire polished.   The diastereoisomer L-*cis*-$[Co(en_2)(NO_2)_2]$ (d-$SbOC_4H_4O_6$) will precipitate in the form of minute, yellow crystals. Allow the solution to stand with intermittent shaking for exactly 10 min. at 25°C. and then quickly filter it through a Büchner funnel under suction.   Retain the filtrate (I) for later recovery of the *dextro* enantiomer.   Wash the crystals of diastereosiomer on the Büchner funnel with 20 ml. portions of 50% ethanol–water, 100% ethanol, and acetone in that order, and allow the crystals to dry in air at room temperature. The yield is 5.5 g. of diastereoisomer, $[a]_D^{24°} = +47.4°$.

Grind 5.5 g. of the diastereoisomer with 30 ml. of water in a glass mortar with a glass pestle, add 7.0 g. of NaI and triturate the mixture for 2 min.   Filter off the slightly soluble L-*cis*-$[Co(en)_2(NO_2)_2]I$ on a Büchner funnel and wash the filtrate once with 15 ml. of water at 0°C. Transfer the precipitate and the filter paper into a 100 ml. Erlenmeyer flask, add 30 ml. of water at 55°C., and shake vigorously.   Add 4.0 g. of freshly precipitated silver chloride and continue the vigorous shaking for 3 min.   Filter off the silver iodide on a Büchner funnel and wash the precipitate with 5 ml. of water at 55°C., allowing the wash water to become part of the filtrate.   To the filtrate add 3.0 g. of ammonium bromide in an ice bath.   Scratch the inside of the flask beneath the surface of the liquid with a stirring rod having an end which is not fire-polished in order to induce crystallization of L-*cis*-$[Co(en)_2(NO_2)_2]Br$.   Filter the product through a Büchner funnel, wash it with 10 ml. portions of 50% ethanol–water, 100% ethanol,

TABLE II

Optical Isomers of Some Coordination Compounds

Metal ion	Complex	References	Ligand
Al(III)	$[Al(TS_2)]^+$	(91)	TS_2 = bis(salicylaldehyde)triethylenetetramine dinegative anion
As(V)	$K[As(cat)_3] \cdot H_2O$	(92)	cat = catechol dinegative anion
Be(II)	$[Be(bzac)_2]$	(46c)	bzac = benzoylacetonate anion
B(III)	$[B(sal)_2]^-$	(93)	sal = salicylate dinegative anion
Cd(II)	$[Cd(en)_3]Cl_2$	(92)	en = ethylenediamine
Cr(III)	$[Cr(en)_3]^{3+}$	(94)	
	$cis\text{-}[Cr(en)_2Cl_2]^+$	(95)	
	$cis\text{-}[Cr(l\text{-}pn)_2Cl_2]^+$	(96)	l-pn = levo-propylenediamine
	$K_3[Cr(C_2O_4)_3]$	(21,65,97)	
	$K[Cr(en)(C_2O_4)_2] \cdot 2H_2O$	(97)	
	$[Cr(acac)_3]$	(51a)	acac = acetylacetonate anion
Co(II)	$[Co(l\text{-}pn)_3]^{2+}$	(98)	
	$[Co(d\text{-}glut)_2]$	(99)	d-glut = d-glutamate mononegative anion
	$[Co(pdta)]^{2-}$	(71)	pdta = propylenediaminetetraacetate tetranegative anion
Co(III)	$[Co(en)_3]^{3+}$	(1,62,68, 100,101)	
	$cis\text{-}[Co(en)_2(NH_3)_2]^{3+}$	(102,103)	
	$cis\text{-}[Co(en)_2(H_2O)_2]^{3+}$	(104)	
	$cis\text{-}[Co(en)_2(NH_3)(NO_2)]^{2+}$	(105)	
	$cis\text{-}[Co(en)_2Cl_2]^+$	(106,107)	
	$[Co(en)_2CO_3]^+$	(108)	
	$[Co(en)(C_2O_4)_2]^-$	(109)	
	$[Co(pn)_3]^{3+}$	(40,76,110)	
	$cis\text{-}[Co(l\text{-}pn)_2(NO_2)_2]^+$	(111)	
	$[Co(glyc)_3]$	(48a)	glyc = glycinate anion
	$[Co(edta)]^-$	(35,39,40, 83,112, 113)	edta = ethylenediaminetetraacetate tetranegative anion
	$[Co(acac)_3]$	(26a,51a, 62)	
	$[Co(C_2O_4)_3]^{3-}$	(34,114, 115)	
Co(III)– Co(III) (binuclear)	$[(en)_2Co(NH_2)(OH)Co(en)_2]^{4+}$	(116)	
Co(III) (Polynuclear)	$[Co\{(OH)_2Co(NH_3)_4\}_3]^{6+}$	(117)	
	$[Co\{(OH)_2Co(en)_2\}_3]^{6+}$	(118)	

(continued)

TABLE II (*continued*)

Metal ion	Complex	References	Ligand
Cu(II)	[Cu(TS$_2$)]	(91)	
Ge(IV)	[Ge(C$_2$O$_4$)$_3$]$^{2-}$	(119)	
Ir(III)	[Ir(en)$_3$]$^{3+}$	(120)	
	[Ir(C$_2$O$_4$)$_3$]$^{3-}$	(121)	
Fe(II)	[Fe(dipy)$_3$]$^{2+}$	(122)	dipy = 2,2′-dipyridyl
	[Fe(*o*-phen)$_3$]$^{2+}$	(66)	*o*-phen = *ortho*-phenanthroline
Fe(III)	[Fe(dipy)$_3$]$^{3+}$	(123)	
	Fe[(*o*-phen)$_3$]$^{3+}$	(66)	
Ni(II)	[Ni(dipy)$_3$]$^{2+}$	(124)	
	[Ni(*o*-phen)$_3$]$^{2+}$	(125)	
Os(II)	[Os(dipy)$_3$]$^{2+}$	(69b,126, 127)	
	[Os(*o*-phen)$_3$]$^{2+}$	(128)	
Os(III)	[Os(dipy)$_3$]$^{3+}$	(69b,126)	
	[Os(*o*-phen)$_3$]$^{3+}$	(123)	
Pd	[Pd(*i*-bn)(*m*-stien)]$^{2+}$	(129)	*i*-bn = *iso*-butylenediamine
	[Pd(*l*-pn)$_2$]$^{2+}$	(130)	*m*-stien = *meso*-stilbenediamine
Pt(II)	[Pt(*i*-bn)(*m*-stien)]$^{2+}$	(2)	
	[Pt(en)(*l*-pn)]$^{2+}$	(130)	
Pt(IV)	[Pt(en)$_3$]$^{4+}$	(131)	
	[Pt(pn)$_3$]$^{4+}$	(77)	
	trans-[Pt(*i*-bn)(*m*-stein)-Cl$_2$]$^{2+}$	(79)	
Rh(III)	[Rh(en)$_3$]$^{3+}$	(132)	
	cis-[Rh(en)$_2$Cl$_2$]$^+$	(133)	
	[Rh(C$_2$O$_4$)$_3$]$^{3-}$	(134)	
	[Rh(NHSO$_2$NH)$_2$-(H$_2$O)$_2$]$^-$	(135)	
Ru(II)	[Ru(dipy)$_3$]$^{2+}$	(69a,136)	
	[Ru(*o*-phen)$_3$]$^{2+}$	(69a,137)	
Ru(III)	[Ru(dipy)$_3$]$^{3+}$	(69,137)	
	[Ru(*o*-phen)$_3$]$^{3+}$	(137,138)	
Si(IV)	[Si(acac)$_3$]$^+$	(3,4)	
Ti(IV)	[Ti(cat)$_3$]$^{2-}$	(139)	
Zn(II)	[Zn(en)$_3$]$^{2+}$	(140)	
	[Zn(qsa)$_2$]	(36)	qsa = 8-quinolino-5-sulfonic acid anion

and acetone in that order and allow it to dry in air. The yield is 3.1 g. (47%) of a product for which $[\alpha]_D = -44°$.

To isolate the *dextro* enantiomer add 2.0 g. of ammonium bromide to the filtrate (I) immediately after removing the *levo* isomer from it, scratch the sides of the flask with a glass rod as before, and allow the impure *dextro* enantiomer to crystallize as the bromide for 5 min. at

20–25°C. Transfer the precipitate on the filter paper into a 100 ml. Erlenmeyer flask, add 35 ml. of water at 55°C. and shake vigorously. Add 4.0 g. of freshly precipitated silver chloride and continue shaking the flask vigorously for another 3 min. Filter off the precipitated silver iodide on a Büchner funnel, wash the precipitate with 5 ml. of water at 55°C., and allow the wash water to become part of the filtrate. To the filtrate add 1.5 g. of NH_4Br, stir, and keep the temperature of the solution at 20–25°C. for 10 min. The D-cis-$[Co(en)_2(NO_2)_2]Br$ crystallizes and should be collected, washed, and dried as described above for the *levo* salt. The yield is 3.1 g. (47%) and $[\alpha]_D = +44°$.

VI. SUMMARY

Optical isomers of over two hundred coordination compounds are known and several compilations of such isomers appear in the literature (84b–89). A comprehensive listing has been prepared by Woldbye (90). Some complexes of interest for which enantiomers have been prepared by one or more of the techniques described in this chapter are listed in Table II along with the appropriate references.

REFERENCES

1. Werner, A., *Chem. Ber.* **45**, 121 (1912).
2. Mills, W. H., and T. H. H. Quibell, *J. Chem. Soc.* **1935**, 839.
3. Dhar, S. K., V. F. Doron, and S. Kirschner, *J. Am. Chem. Soc.* **80**, 753 (1958).
4. Dhar, S. K., V. F. Doron, and S. Kirschner, *J. Am. Chem. Soc.* **81**, 6372 (1959).
5. Mathieu, J. P., *Bull. Soc. Chim. France* **6**, 873, 1258 (1938).
6. Jaeger, F. M., *Optical Activity and High Temperature Measurements*, McGraw-Hill, New York, 1930.
7. Matoush, W. R., and F. Basolo, *J. Am. Chem. Soc.* **78**, 3972 (1956).
8. Albinak, M. J., D. C. Bhatnagar, S. Kirschner and A. J. Sonnessa, *Can. J. Chem.* **39**, 2360 (1961); *Advances in the Chemistry of the Coordination Compounds*, S. Kirschner, ed., Macmillan, New York, 1961, pp. 154 ff.
9. Brushmiller, J. G., E. L. Amma, and B. E. Douglas, *J. Am. Chem. Soc.* **84**, 111 (1962).
10. Collman, J. P., R. P. Blair, R. L. Marshall, and S. Slade, *Inorg. Chem.* **2**, 576 (1963).
11. Kirschner, S., *J. Am. Chem. Soc.* **78**, 2372 (1956); A. Luttringhaus and D. Berrer, *Tetrahedron Letters* **1959**, No 10, p. 10.
12. Arago, E., *Memoir de l'Inst.* **12**, Part I, 93 (1811).
13. Biot, J. B., *Memoir de l'Inst.* **13**, Part I, 1 (1812).
14. Heller, W., and D. Fitts, in *Physical Methods of Organic Chemistry*, 3rd ed., A. Weissberger, ed., of the Technique of Organic Chemistry Series, Vol. I, Part 3, Interscience, New York, 1960, Chap. 33.
15. Lowry, T. M., *Optical Rotatory Power*, Longmans, Green, London, 1935.
16. Moscowitz, A., in *Optical Rotatory Dispersion*, C. Djerassi, ed., McGraw-Hill, New York, 1960, Chap. 12.

COORDINATION COMPOUNDS 53

17. Djerassi, C., *Optical Rotatory Dispersion*, McGraw-Hill, New York, 1960; W. Kauzmann, J. Walter, and H. Eyring, *Chem. Rev.* **26**, 339 (1940); E. Condon, *Rev. Mod. Phys.* **9**, 432 (1937); W. Kuhn, *Ann. Rev. Phys. Chem.* **9**, 417 (1958); W. Moffitt, *J. Chem. Phys.* **25**, 1189 (1956); H. M. Powell, *Endeavour*, **15**, 20 (1956); M. Vol'kenshtein, *Zh. Eksperim. Teor. Fiz.* **20**, 342 (1950); R. Servant, *J. Phys. Radium* **3**, 90 (1942); J. Kirkwood, *J. Chem. Phys.* **5**, 479 (1937); M. Born, *Proc. Roy. Soc. (London)* **A150**, 84 (1935); **A153**, 339 (1936); M. Betti, *Trans. Faraday Soc.* **26**, 337 (1930); F. Woldbye, *Acta Chem. Scand.* **13**, 2137 (1959).
18. Moeller, T., *Inorganic Chemistry*, Wiley, New York, 1952, p. 265.
19. Wilkins, R. G., and M. J. G. Williams in *Modern Coordination Chemistry*, J. Lewis and R. G. Wilkins, eds., Interscience, New York, 1960, p. 202.
20. Pasteur, L., *Ann. Chim. Phys.* **24**, 442 (1848).
21. Jaeger, F. M., *Rec. Trav. Chim.* **38**, 250 (1919).
22a. Werner, A., and J. Bosshart, *Chem. Ber.* **47**, 2171 (1914); b. R. M. Secor, *Chem. Rev.* **63**, 297 (1963).
23. Delépine, M., *Bull. Soc. Chim. France* **29**, 656 (1921); **1**, 1256 (1934).
24a. Delépine, M., and R. Charonnat, *Bull. Soc. Franc. Mineral.* **53**, 73 (1930); b. M. M. Jamison and E. E. Turner, *J. Chem. Soc.* **1942**, 437.
25. Dwyer, F. P., E. C. Gyarfas, and M. F. O'Dwyer, *Nature* **167**, 1036 (1951).
26. Dwyer, F. P., and E. C. Gyarfas, *Nature* **168**, 29 (1951); N. R. Davies and F. P. Dwyer, *Trans. Faraday Soc.* **50**, 24 (1954).
27. Pasteur, L., *Compt. Rend.* **46**, 615 (1858).
28. Price, V. E., and J. P. Greenstein, *J. Biol. Chem.* **175**, 969 (1948).
29. Greenstein, J. P., "The Resolution of Racemic α-Amino-Acids," in *Advances in Protein Chemistry*, Vol. 9, M. L. Anson, K. Bailey, and J. T. Edsall, eds., Academic Press, New York, 1954, pp. 121-202.
30. Crespi, M. J. S., *Dissertation Abstr.* **20**, 4512 (1960).
31. Bailey, M. E., and H. B. Haas, *J. Am. Chem. Soc.* **63**, 1969 (1941).
32. Butler, C. L., and L. H. Cretcher, *J. Am. Chem. Soc.* **55**, 2605 (1933).
33. Zetzsche, F., and M. Hubacher, *Helv. Chim. Acta* **9**, 291 (1926).
34. Dwyer, F. P., and A. M. Sargeson, *J. Phys. Chem.* **60**, 1331 (1956).
35. Dwyer, F. P., and F. L. Garvan, in *Inorganic Syntheses* Vol. 6, E. G. Rochow, ed., McGraw-Hill, New York, 1960, p. 195.
36. Liu, J. C. I., and J. C. Bailar, Jr., *J. Am. Chem. Soc.* **73**, 5432 (1951).
37. Bailar, J. C., Jr., H. B. Jonassen, and A. D. Gott, *J. Am. Chem. Soc.* **74**, 3131 (1952).
38. Gott, A. D., and J. C. Bailar, Jr., *J. Am. Chem. Soc.* **74**, 4820 (1952).
39. Dwyer, F. P., E. C. Gyarfas, and D. P. Mellor, *J. Phys. Chem.* **59**, 296 (1955).
40. Kirschner, S., Y. K. Wei, and J. C. Bailar, Jr., *J. Am. Chem. Soc.* **79**, 5877 (1957).
41. Das Sarma, B., and J. C. Bailar, Jr., *J. Am. Chem. Soc.* **77**, 5480 (1955).
42. Kuhn, W., and E. Braun, *Naturwissenschaften* **17**, 227 (1929).
43. Kuhn, W., and E. Knopf, *Z. Physik. Chem. (Leipzig)* **7**, 292 (1930); *Naturwissenschaften* **18**, 198 (1930).
44a. Jaeger, F. M., and G. Berger, *Rec. Trav. Chim.* **40**, 153 (1921); b. R. Tsuchida, A. Nakamura, and M. Kobayashi, *J. Chem. Soc. Japan* **56**, 1335 (1935).
45. Tsuchida, R., M. Kobayashi, and A. Nakamura, *Bull. Chem. Soc. Japan* **11**, 38 (1936); H. Kuroya and R. Tsuchida, *Bull. Chem. Soc. Japan* **15**, 427 (1940); A. Nakahara and R. Tsuchida, *J. Am. Chem. Soc.* **76**, 3103 (1954).

46a. Bailar, J. C., Jr., and D. F. Peppard, *J. Am. Chem. Soc.* **62**, 105 (1940); b. D. H. Busch and J. C. Bailar, Jr., *J. Am. Chem. Soc.* **75**, 4574 (1953); **76**, 5352 (1954); c. H. Irving and J. B. Gill, *Proc. Chem. Soc.* **1958**, 168; d. J. R. Kuebler, Jr. and J. C. Bailar, Jr., *J. Am. Chem. Soc.* **74**, 3535 (1952); e. B. Das Sarma and J. C. Bailar, Jr., *J. Am. Chem. Soc.* **77**, 5476 (1955); f. G. K. Schweitzer and C. K. Talbot, *J. Tenn. Acad. Sci.* **25**, 143 (1950).

47. Ferroni, E., and R. Cini, *J. Am. Chem. Soc.* **82**, 2427 (1960).

48a. Krebs, H., and R. Rasche, *Z. Anorg. Allgem. Chem.* **276**, 236 (1954); *Naturwissenschaften* **41**, 63 (1954); b. H. Krebs, J. Diewald, and J. Wagner, *Angew. Chem.* **67**, 705 (1955); c. H. Krebs, J. Diewald, H. Arlitt, and J. Wagner, *Z. Anorg. Allgem. Chem.* **287**, 98 (1956); d. H. Krebs, J. Diewald, R. Rasche, and J. Wagner, *Die Trennung von Racematen auf Chromatographischen Wege*, Westdeutscher Verlag, Cologne, 1956.

49. Karagounis, G., E. Charbonnier, and E. Flöss, *J. Chromatog.* **2**, 84 (1959).

50. Piper, T. S., *J. Am. Chem. Soc.* **83**, 3908 (1961).

51a. Moeller, T., and E. Gulyas, *J. Inorg. Nucl. Chem.* **5**, 245 (1958); **9**, 82 (1958); b. T. M. Hseu, D. F. Martin, and T. Moeller, *Inorg. Chem.* **2**, 587 (1963).

52. Mayer, W., and F. Merger, *Ann. Chem.* **644**, 65 (1961).

53. Arnstein, H. R. V., and D. Morris, *Biochem. J.* **76**, 318 (1960); M. Jaeger, S. Iskric, and M. Wickerhauser, *Croat. Chem. Acta* **28**, 5 (1956).

54. Tsuboyama, S., and M. Yanagita, *Sci. Papers Inst. Phys. Chem. Res. (Tokyo)* **53**, 245 (1959).

55. Buchar, E., and K. Suchy, *Magy. Kem. Folyoirat* **64**, 45 (1958).

56. Kikkawa, I., *Yakugaku Zasshi* **81**, 732 (1961).

57. Kotake, M., T. Sakan, N. Nakamura, and S. Senoh, *J. Am. Chem. Soc.* **73**, 2973 (1951).

58. Curti, R., and U. Colombo, *J. Am. Chem. Soc.* **74**, 3961 (1952); *Gazz. Chim. Ital.* **82**, 491 (1952).

59. Klemm, L. H., and D. Reed, *J. Chromatog.* **3**, 634 (1960).

60. Carassiti, V., *Ann. Chim.* **47**, 1337 (1957); **48**, 167 (1958); *J. Chim. Phys.* **55**, 120 (1958).

61. Jamison, M. M., and E. E. Turner, *J. Chem. Soc.* **1942**, 611; E. Gil-Av, and D. Nurok, *Proc. Chem. Soc.* **1962**, 146.

62. Doron, V. F., and S. Kirschner, *Inorg. Chem.* **1**, 539 (1962).

63. Karagounis, G., and G. Lippold, *Naturwissenschaften* **46**, 145 (1959).

64. Sievers, R., R. W. Moshier, and M. L. Morris, *Inorg. Chem.* **1**, 966 (1962); G. Goldberg and W. A. Ross, *Chem. Ind. (London)* **1962**, 657.

65. Werner, A., *Chem. Ber.* **45**, 3061 (1912).

66. Dwyer, F. P., and E. C. Gyarfas, *J. Proc. Roy. Soc., N.S. Wales* **83**, 263 (1950).

67. Jonassen, H. B., J. C. Bailar, Jr., and E. H. Huffman, *J. Am. Chem. Soc.* **70**, 756 (1948).

68. Busch, D. H., *J. Am. Chem. Soc.* **77**, 2747 (1955).

69a. Dwyer, F. P., and E. C. Gyarfas, *J. Proc. Roy. Soc. N.S. Wales* **83**, 170, 174 (1949); b. *Nature* **166**, 481 (1950).

70. Brandt, W. W., F. P. Dwyer, and E. C. Gyarfas, *Chem. Rev.* **54**, 997 (1954).

71. Im, Y. A., and D. H. Busch, *J. Am. Chem. Soc.* **83**, 3362 (1961); D. H. Busch, D. W. Cooke, K. Swaminathan, and Y. A. Im, in *Advances in the Chemistry of the Coordination Compounds*, S. Kirschner, ed., Macmillan, New York, 1961, p. 139.

72. Eliel, E. L., *Stereochemistry of Carbon Compounds*, McGraw-Hill, New York, 1962.

73. Fischer, E., *Chem. Ber.* **27**, 3189, 3231 (1894).
74. Marckwald, W., *Chem. Ber.* **37**, 1368 (1904).
75. Davis, T. L., and R. Heggie, *J. Am. Chem. Soc.* **57**, 377 (1935).
76. Tschugaeff, L., and W. Sokoloff, *Chem. Ber.* **42**, 55 (1909).
77. Smirnoff, A. P., *Helv. Chim. Acta* **3**, 177 (1920).
78. Jaeger, F. M., and H. B. Blumendal, *Z. Anorg. Allgem. Chem.* **175**, 161 (1928); F. M. Jaeger, *Koninkl. Ned. Akad. Wetenschof. Proc.* **40**, 108 (1937).
79. Messing, A. F., and F. Basolo, *J. Am. Chem. Soc.* **78**, 4511 (1956).
80. Bailar, J. C., Jr. and R. W. Auten, *J. Am. Chem. Soc.* **56**, 774 (1934).
81. Burwell, R. L., A. D. Shields, and H. Hart, *J. Am. Chem. Soc.* **76**, 908 (1954).
82. Berson, J. A., and D. A. Ben-Efraim, *J. Am. Chem. Soc.* **81**, 4083 (1959).
83. Graff, S., D. Rittenberg, and G. L. Koster, *J. Biol. Chem.* **133**, 745 (1940).
84a. Holtzclaw, H. F., D. P. Sheetz, and B. D. McCarty, in *Inorganic Syntheses*, Vol. 4, J. C. Bailar, Jr., ed., McGraw-Hill, New York, 1953, p. 176; b. D. H. Busch, in *Cobalt*, R. S. Young, ed., Reinhold, New York, 1960, p. 94.
85. Basolo, F., *Chem. Rev.* **52**, 459 (1953).
86. Martell, A. E., and M. Calvin, *Chemistry of the Metal Chelate Compounds*, Prentice-Hall, Princeton, New Jersey, 1952.
87. Wilkins, R. G., and M. J. G. Williams, in *Modern Coordination Chemistry*, J. Lewis and R. F. Wilkins, eds., Interscience, New York, 1960, Chap. 3.
88. Basolo, F., in *Chemistry of the Coordination Compounds*, J. C. Bailar, Jr., ed., Reinhold, New York, 1956, Chap. 8.
89. Basolo, F., and R. G. Pearson, *Mechanisms of Inorganic Reactions*, Wiley, New York, 1958, p. 258.
90. Woldbye, F., "Technique of Optical Rotatory Dispersion and Circular Dichroism," in *Technique of Inorganic Chemistry*, Vol. IV, H. B. Jonassen and A. Weissberger, eds., Interscience, New York, to be published.
91. Das Sarma, B., and J. C. Bailar, Jr., *J. Am. Chem. Soc.* **77**, 5476 (1955).
92. Weinland, R. F., and J. Heinzler, *Chem. Ber.* **52**, 1322 (1919); A. Rosenheim and W. Plato, *Chem. Ber.* **58**, 2000 (1925).
93. Faulkner, I. J., and T. M. Lowry, *J. Chem. Soc.* **127**, 1080 (1925).
94. Werner, A., *Chem. Ber.* **45**, 865 (1912).
95. Werner, A., *Chem. Ber.* **44**, 3138 (1911); J. Selbin and J. C. Bailar, Jr., *J. Am. Chem. Soc.* **79**, 4285 (1957).
96. Rollinson, C. L., and J. C. Bailar, Jr., *J. Am. Chem. Soc.* **66**, 641 (1944).
97. Bushra, E., and C. H. Johnson, *J. Chem. Soc.* **1939**, 1937.
98. Woldbye, F., *Optical Rotatory Dispersion of Transition Metal Complexes*, European Research Office, U. S. Army, Frankfurt a.M., 1959; *Record Chem. Progr.* (*Kresge-Hooker Sci. Lib.*) **24**, 197 (1963).
99. Lifschitz, I., and F. L. M. Schouteden, *Rec. Trav. Chim.* **58**, 411 (1939).
100. Douglas, B. E., *J. Am. Chem. Soc.* **76**, 1020 (1954).
101. Broomhead, J. A., F. P. Dwyer, and J. W. Hogarth, in *Inorganic Syntheses*, Vol. 6, E. G. Rochow, ed., McGraw-Hill, New York, 1960, p. 183.
102. Werner, A., and Y. Shibata, *Chem. Ber.* **45**, 3287 (1912).
103. Archer, R., and J. C. Bailar, Jr., *J. Am. Chem. Soc.* **83**, 812 (1961).
104. Mathieu, J. P., *Bull. Soc. Chim. France* **4**, 687 (1937).
105. Mathieu, J. P., *Bull. Soc. Chim. France* **3**, 476, 463 (1936).
106. Werner, A., *Chem. Ber.* **44**, 3279 (1911).
107. Bailar, J. C., Jr., in *Inorganic Syntheses*, Vol. 2, W. C. Fernelius, ed., McGraw-Hill, New York, 1946, p. 222.

108. Werner, A., and M. Cutcheon, *Chem. Ber.* **45**, 3281 (1912).

109. Dwyer, F. P., I. K. Reid, and F. L. Garvan, *J. Am. Chem. Soc.* **83**, 1285 (1961).

110. Dwyer, F. P., F. L. Garvan, and A. Shulman, *J. Am. Chem. Soc.* **81**, 290 (1959).

111. O'Brien, T. D., J. P. McReynolds, and J. C. Bailar, Jr., *J. Am. Chem. Soc.* **70**, 749 (1948).

112. Busch, D. H., and J. C. Bailar, Jr., *J. Am. Chem. Soc.* **75**, 4574 (1953).

113. Weakliem, H. A., and J. L. Hoard, *J. Am. Chem. Soc.* **81**, 549 (1959).

114. Delépine, M., *Bull. Soc. Chim. France* **1**, 1256 (1934).

115. Jaeger, F. M., and W. Thomas, *Koninkl. Ned. Akad. Wetenschof. Proc.* **21**, 693 (1918).

116. Werner, A., *Chem. Ber.* **47**, 1961 (1914).

117. Werner, A., *Chem. Ber.* **47**, 3087 (1914); *Compt. Rend.* **159**, 426 (1914).

118. Goodwin, H. A., E. C. Gyarfas, and D. P. Mellor, *Australian J. Chem.* **11**, 426 (1958).

119. Moeller, T., and N. C. Nielsen, *J. Am. Chem. Soc.* **75**, 5106 (1953).

120. Werner, A., and A. P. Smirnoff, *Helv. Chim. Acta* **3**, 472 (1920); J. P. Mathieu, *J. Chim. Phys.* **33**, 78 (1936).

121. Delépine, M., *Compt. Rend.* **159**, 239 (1914); **175**, 1409 (1922).

122. Werner, A., *Chem. Ber.* **45**, 433 (1911); F. P. Dwyer and E. C. Gyarfas, *J. Proc. Roy. Soc. N.S. Wales* **85**, 126 (1951).

123. Dwyer, F. P., and E. C. Gyarfas, *J. Am. Chem. Soc.* **74**, 4699 (1952).

124. Morgan, G. T., and F. H. Burstall, *J. Chem. Soc.* **1931**, 2213; F. M. Jaeger and J. A. van Dijk, *Z. Anorg. Allgem. Chem.* **227**, 304 (1936); G. K. Schweitzer and J. M. Lee, *J. Phys. Chem.* **56**, 195 (1952).

125. Dwyer, F. P., and E. C. Gyarfas, *J. Proc. Roy. Soc., N.S. Wales* **83**, 232 (1949),

126. Dwyer, F. P., and E. C. Gyarfas, *J. Am. Chem. Soc.* **73**, 2322 (1951).

127. Burstall, F. H., F. P. Dwyer, and E. C. Gyarfas, *J. Chem. Soc.* **1950**, 953.

128. Dwyer, F. P., E. C. Gyarfas, and N. A. Gibson, *J. Proc. Roy. Soc., N.S. Wales* **84**, 68 (1950).

129. Lidstone, A. G., and W. H. Mills, *J. Chem. Soc.* **1939**, 1754.

130. Tchugaeff, L., and W. Sokoloff, *Chem. Ber.* **40**, 3461 (1907).

131. Werner, A., *Vierteljahressch. Naturforsch. Ges. Zuerich* **62**, 553 (1917); J. P. Mathieu, *Bull. Soc. Chim. France* **6**, 1258 (1939); *J. Chim. Phys.* **33**, 28 (1936).

132. Werner, A., *Chem. Ber.* **45**, 1228 (1912).

133. Andersen, S., and F. Basolo, *J. Am. Chem. Soc.* **82**, 5953 (1960).

134. Werner, A., *Chem. Ber.* **45**, 1954 (1914).

135. Mann, F. G., *J. Chem. Soc.* **1933**, 412.

136. Burstall, F. H., *J. Chem. Soc.* **1936**, 173.

137. Dwyer, F. P., and E. C. Gyarfas, *Nature* **163**, 918 (1949).

138. Dwyer, F. P., and E. C. Gyarfas, *J. Proc. Roy. Soc., N.S. Wales* **83**, 170 (1949).

139. Rosenheim, A., B. Raibmann, and G. Schendel, *Z. Anorg. Allgem. Chem.* **196**, 168 (1931).

140. Neogi, P., and G. K. Mukerjee, *J. Indian Chem. Soc.* **11**, 681 (1934).

SUGGESTED BIBLIOGRAPHY

Bailar, J. C., Jr., ed., *Chemistry of the Coordination Compounds*, Reinhold, New York, 1956.

Basolo, F., and R. G. Pearson, *Mechanisms of Inorganic Reactions*, Wiley, New York, 1958.

Bates, F. J., et al., *Polarimetry, Saccharimetry and the Sugars*, U. S. Bureau of Standards, Circular C440, U. S. Government Printing Office, Washington, D. C., 1942.

de la Mare, P. B. D., and W. Klyne, eds., *Progress in Stereochemistry*, Vol. III, Butterworths, London, 1962.

Djerassi, C., *Optical Rotatory Dispersion*, McGraw-Hill, New York. 1960.

Eliel, E. L., *Stereochemistry of Carbon Compounds*, McGraw-Hill, New York, 1962.

Fernelius, W. C., ed., *Inorganic Syntheses*, Vol. 2, McGraw-Hill, New York, 1946.

Jaeger, F. M., *Optical Activity and High Temperature Measurements*, McGraw-Hill, New York, 1930.

Kirschner, S., ed., *Advances in the Chemistry of the Coordination Compounds*, Macmillan, New York, 1961.

Klyne, W., and P. B. D. de la Mare, eds., *Progress in Stereochemistry*, Vol. II, Butterworths, London, 1958.

Klyne, W., ed., *Progress in Stereochemistry*, Vol. I., Academic Press, New York, 1954.

Lewis, J., and R. G. Wilkins, eds., *Modern Coordination Chemistry*, Interscience, New York, 1960.

Lowry, T. M., *Optical Rotatory Power*, Longmans, Green, London, 1935.

Martell, A. E., and M. Calvin, *Chemistry of the Metal Chelate Compounds*, Prentice-Hall, Englewood Cliffs, New Jersey, 1952.

Rochow, E. G., ed., *Inorganic Syntheses*, Vol. 6, McGraw-Hill, New York, 1960.

Weissberger, A., ed., *Physical Methods of Organic Chemistry*, 3rd Ed., Vol. I, Part 3, of Technique of Inorganic Chemistry Series, Interscience, New York, 1960.

Wells, A. F., *Structural Inorganic Chemistry*, Clarendon, Oxford, 1962.

Werner, A., *Neure Anschauungen auf dem Gebiete der Anorganischen Chemie*, F. Vieweg and Sohn, Verlag, Braunschweig, Germany, 1920.

Woldbye, F., "Technique of Optical Rotatory Dispersion and Circular Dichroism," in *Technique of Inorganic Chemistry*, Vol. IV, H. B. Jonassen and A. Weissberger eds., Interscience, New York, to be published.

Young, R. S., *Cobalt*, Reinhold, New York, 1960.

CHAPTER 3

Metal Derivatives of Ketimine and Aldimine Compounds

DEAN F. MARTIN

University of Illinois, Urbana, Illinois

CONTENTS

I. INTRODUCTION

When an amine undergoes a condensation reaction with a carbonyl compound a Schiff base, $RR'C = NR''$, forms. The Schiff base may be an aldimine or a ketimine, depending upon whether the carbonyl compound is an aldehyde or a ketone, respectively,

$$RCHO + R''NH_2 \rightleftharpoons RCH = NR'' + H_2O$$

$$RR'CO + R''NH_2 \rightleftharpoons RR'C = NR'' + H_2O$$

59

These products have received considerable attention as model compounds for theoretical studies and as precursors to heterocyclic compounds. Schiff bases are of especial interest because of their ability to form coordination compounds. These coordination compounds have stimulated work in physical chemistry (magnetochemistry, infrared and ultraviolet absorption spectra, dipole moments, etc.), studies of stereochemistry, coordination polymers, and mechanisms of inorganic reactions. Because of the interest in the metal derivatives of aldimines and ketimines, it seems worthwhile to review the methods of preparation.

A. Nomenclature

It seems appropriate to consider briefly the nomenclature that will be used in this review.

Frequently, names will be used for the condensation products of amines and carbonyl compounds. Thus, $[CH_3COCH_2C(CH_3)=NCH_2]_2$ obtained from the condensation of acetylacetone and ethylenediamine, will be called bis-acetylacetoneethylenediamine, though the systematic (I.U.C.) name is 4,4'-(ethylenedinitrilo)dipentane-2-one or 4,4'-(ethylenediamino)di-2-pentene-2-one. Similar practice will be followed for naming Schiff bases derived from substituted salicylaldehydes. Systematic nomenclature will be applied to β-ketoimines of the type $RCOCH_2C(=NR)R''$. Thus, $CH_3COCH_2C(=NH)CH_3$ will be named 4-iminopentane-2-one and the copper(II) derivative, bis(4-iminopentane-2-ono)copper(II). It should be noted that many workers prefer to indicate that the ligand exists in the ketoamine form, $CH_3COCH=C(NH_2)CH_3$; in this case, the ligand and copper(II) derivative would be named 4-amino-3-pentene-2-one and bis(4-amino-3-pentene-2-ono)-copper(II), respectively.

B. Side-Reactions

The reaction of ketimines and aldimines with metal ions may be accompanied by certain side-reactions. These include hydrolysis, amine exchange, and cleavage reactions.

Hydrolysis is a likely side-reaction inasmuch as this represents a reverse of the formation of the Schiff base:

$$\diagup C=N \diagdown R \quad + H_2O \rightarrow \quad \diagup C=O \quad + RNH_2$$

Schiff bases vary in susceptibility to hydrolysis: bis-acetylacetoneethylenediamine, $[CH_3COCH_2C(CH_3)NCH_2]_2$, can be recrystallized from water (1), but the tetramethylene analog is unstable in water (2). Because of the basicity of the amine, the presence of acids will cause the hydrolysis reaction to be favored in the Schiff base formation–hydrolysis equilibrium. Since hydrogen ion may be liberated by a Schiff base, HCh, upon coordination with a metal ion,

$$n\text{HCh} + \text{M}^{n+} \rightarrow \text{MCh}_n + n\text{H}^+$$

it is necessary to control the pH of the reaction solution.

Depending upon the ligand, the act of coordination may stabilize or labilize the Schiff base with regard to hydrolysis. Interestingly, the susceptibility to hydrolysis has been used to advantage to provide a homogeneous system for the synthesis of metal-β-diketone compounds (3):

$$2\text{RCOCH}{=}\text{C(R}''\text{)NR}'_2 + \text{MX}_2 \xrightarrow{\text{H}_2\text{O}} (\text{RCOCH}{=}\text{COR}'')_2\text{M} + 2\text{R}'_2\text{NH}_2\text{X}$$

Often hydrolysis can be prevented by control of the pH by the use of a buffer system or by addition of a weak base such as ammonia. If the latter is used, amine exchange may occur. For example, the interaction of bis-acetylacetonetrimethylenediamine and copper ion in the presence of a large excess of ammonia results (4) in formation of bis(4-iminopentane-2-one)copper(II):

$$[\text{H}_3\text{CCOCHC}({=}\text{NH})\text{CH}_3]_2\text{Cu} + \text{H}_3\text{N(CH}_2)_3\text{NH}_3{}^{2+}$$

Cleavage of the N—R bond in Schiff-base compounds has also been observed. Pfeiffer and co-workers (5) reported the failure to prepare compounds of the type illustrated; the only crystalline product isolated

was bis-salicylaldiminonickel(II). Oxidative cleavage was suggested because the reaction did not occur in the absence of air. Similarly, an attempt to prepare [CH$_3$COCHC($=$NCH$_3$)CH$_3$]$_2$Cu by interaction of the ligand, copper acetate, in the presence of added acetate ion resulted in the cleavage product $[CH_3COCHC(=NH)CH_3]_2Cu$ (4).

C. Preparation of Ligands

Generally, the salicylaldimine ligands have not been isolated, but several workers (6,7) have isolated such ligands (usually prepared by direct combination of aldehyde and amine, with or without the use of solvent) and have reported physical constants. Occasionally, it is not possible to prepare the ligand by direct combination. For example, the interaction of salicylaldehyde and triethylenetetramine did not afford (8) the desired Schiff base but rather

$$HOC_6H_4CH=N(CH_2)_2N-CH_2CH_2-N(CH_2)_2N=CHC_6H_4OH$$
$$CH$$
$$C_6H_4OH$$

The preparation of β-ketoimines is usually described in connection with the preparation of the metal derivatives (10,11). Cromwell (9) has discussed the methods of preparation and has summarized the pertinent literature.

II. METHODS OF PREPARATION OF METAL DERIVATIVES

A. By Direct Reaction

Two important aspects of the direct reaction of an imine ligand and a metal ion are the choice of solvent and the choice of metal ions. It is worthwhile to consider these separately in terms of the procedures that have been used and the advantages and disadvantages of a given choice.

1. Choice of Solvent

The reaction of an imine and a metal salt in water is generally limited by the slight solubility of most imines in water and is often precluded by the ease with which the imines undergo hydrolysis. There are obvious and important exceptions to this, however. If the imine is mono or multiprotic, it may be soluble in alkaline solution, and, under these conditions, the rate of hydrolysis is reduced greatly. For example, bis-acetylacetoneethylenediamine is soluble in water, and the cobalt(II) compound has been prepared by direct reaction in aqueous solution using an equivalent amount of sodium hydroxide (12). Also, bis(N,N'-disalicyl-ethylenediimine)-μ-aquodicobalt(II) can be prepared by direct reaction in sodium acetate buffered solution (13).

Direct reaction in aqueous solution may also be possible if the imine contains solubilizing groups, as is the case with certain terdentate ligands (ter) derived from pyridine-2-aldehyde and 2-aminomethylpyridine (14). Compounds of the type $[Fe(ter)_2]^{2+}$ are formed (14) when the ligand is added to aqueous iron(II) sulfate; the perchlorate salt is precipitated upon the addition of sodium perchlorate.

It is common practice to avoid the problems of solubility and hydrolysis inherent in aqueous systems and to use a mixed-solvent system. A number of copper salts of β-ketoimines have been prepared by mixing a solution of the ligand in alcohol with ammoniacal copper acetate or nitrate (4,14,15). A similar procedure has been used to prepare bis(4-iminopentane-2-ono)nickel(II) (16,18). The use of ammoniacal solution prevents hydrolysis and shifts the chelation equilibrium. The practice has the disadvantage that extraneous ions are added, and, if the concentration of ammonia is sufficiently great, amine exchange may occur. This problem can be avoided by reducing the amount of ammonium hydroxide used. The interaction of $CH_3COCH_2C(=NCH_3)CH_3$ and copper nitrate results in the expected product when a stoichiometric amount of ammonium hydroxide is used (19), but the amine-exchange product, bis(4-iminopentane-2-ono)copper(II), is obtained when a large excess of ammonia is used (4).

The use of a non-aqueous solvent, in which both the ligand and metal compound are soluble, can eliminate hydrolysis, contamination by extraneous ions, and unfavorable equilibria. Thus, the reaction of metal acetates and β-ketoimines has been effected by heating a dioxane (1), ethanol (2,20), or methanol (21) solution. The product may precipitate on cooling, or it may be precipitated by addition of water.

2. Choice of Metal Compound

In general, metal salts (acetate, nitrate, halide, sulfate) have been used as the source of the metal ion. The use of an acetate salt offers some advantage in the control of pH since an acetate buffer is established.

A number of β-ketoimine-metal chelate compounds have been prepared by refluxing an acetone solution of the ligand over freshly prepared divalent-metal hydroxide (22). The method has been used to prepare copper(II), nickel(II), and cobalt(II) derivatives, but not the zinc(II) or cadmium(II) compounds (22). The reaction may be slow because of the heterogeneous system that results. A disadvantage is that occasionally the product precipitates on the hydroxide and makes further reaction difficult. Usually, the product is soluble in acetone and easy separation from unreacted metal hydroxide is possible.

A review (23) of the methods of preparing metal derivatives of 1,3-diketones suggests other metal compounds that might be used. For example, many metal halides (such as $GeCl_4$, $ZrCl_4$, $ThCl_4$, BCl_3, and $AlCl_3$) are soluble in inert solvents and reaction with a protic ligand would produce insoluble hydrogen halide. Such a reaction seems never to have been reported. However, a solvate, trichloro-tris(tetrahydrofuran)chromium(III), has been used to prepare compounds of the type $Cr[RCOCHC(=NR')R'']_3$ by addition of the solvate to an equimolar mixture of potassium and ligand in *tert*-butanol (10).

B. Constituent Combination

This method consists in mixing, in a suitable order, amine, aldehyde or ketone, base, and metal ion in a suitable solvent. The method is a convenient one since it eliminates the need for preparation of the ligand. The approach may be necessary when a ligand is too unstable to be isolated and purified or when a ligand undergoes undesired condensations in the absence of metal ions. The method has the obvious disadvantage that there may be contamination of the product if one of the reactants is in excess, as is usually the case. Thus, preparation of bis(N,N'-disalicylethylenediimino) μ-aquodicobalt(II) by direct reaction affords a better product than if constituent combination is used (13).

It appears that the order of mixing of reactants is critical. Nunez and Eichhorn (24) have studied the rate of the formation of compounds of the type

in 50% (v/v) dioxane–water solution. The formation of the complex was retarded if copper or nickel ions were allowed to react with one of the Schiff base components (salicylaldehyde or glycine). However, addition of copper or nickel ions to a mixture of Schiff base components resulted in immediate formation of the metal–Schiff base compound, though the concentration of the metal complex was essentially that present at equilibrium. It appears that in this case the Schiff base is formed without participation of metal ions.

Constituent combination has not been used extensively to prepare metal derivatives of β-ketoimines, although the reaction is a potentially useful one. This apparent oversight may be due to the fact that metal derivatives of β-diketones are more insoluble and stable than those of

β-ketoimines. The preparation of bis-acetylacetoneethylenediimino-copper(II) from acetylacetone, ethylenediamine, and copper ion is indicated (25). However, attempts to effect condensation of hexafluoro-acetylacetone and ethylenediamine in the presence of copper acetate were unsuccessful (1).

Two general procedures have been used to prepare salicylaldimine–metal compounds. The Pfeiffer procedure (26) consists in refluxing a methanol solution of bis(salicylaldehyde)copper(II), sodium acetate, and amine; the reactants are in $1:1:1$ ratio by weight. The procedure used by Charles (27,28) consists in adding an aqueous metal acetate solution to a methanol or dilute methanol solution of salicylaldehyde, amine, and sodium acetate, the resulting mixture is refluxed. Verter and Frost (29) report that salicylaldimines derived from aniline are best prepared by Charles' method and that the amino ester derivatives are readily prepared by Pfeiffer's method.

The preparation of tris(biacetyl-bis-methylimine)cobalt(II) and related compounds is a good example of the use of constituent combination to prepare derivatives of neutral imines (30).

Compounds of the type $[o\text{-}OC_6H_4C(CH_3)\text{=}NR]_2M$ have been prepared by mixing the constituents in solution or by the action of ammonia (R=H) on bis(o-hydroxyacetophenono)nickel(II) (31–33).

C. Amine Exchange

This method consists of the displacement of an RN moiety of an imine by an amine

$$\begin{array}{ccc} & M & M \\ & \uparrow & \uparrow \\ \diagdown C\text{=}NR + R'NH_2 \rightarrow & \diagdown C\text{=}NR' + RNH_2 \\ \diagup & \diagup & \end{array}$$

The method has been used by Verter and Frost (29) and by Muto (34) to prepare a variety of compounds of the type

$$[o\text{-}OC_6H_4CH\text{=}NR]_2Cu$$

The procedure is simple: The ligand or copper-chelate compound is suspended in the appropriate amine and the mixture heated; often the product can be isolated by diluting the reaction mixture with water. If the amine is a solid, a solvent (ethanol, chloroform, benzene, toluene, ether, or acetone) may be used (34). Thus, amine exchange offers the advantage that a variety of metal–salicylaldimine compounds can be prepared from one chelate by combining it with appropriate amines. The reaction is conducted under reducing conditions, and the oxidation state of the metal may be reduced.

Although salicylaldimines and the copper(II) derivatives readily undergo amine exchange, the same is not true of β-ketoimines and the copper(II) derivatives. Bis(4-iminopentane-2-ono)copper(II) apparently does not react with ethylenediamine (4), nor was there evidence that 4-iminopentane-2-one reacted with ethylenediamine in 95% ethanol, but this β-ketoimine did undergo amine exchange with ethylenediamine when copper ion was present. This exchange represents a version of the constituent combination procedure.

There has been some speculation (4,29) on the mechanism of amine exchange. An essential feature of the reaction is the polarization of the azomethene linkage through coordination, though this is obviously not a sole requirement because a number of salicylaldimes undergo amine exchange in the absence of metal ion (34).

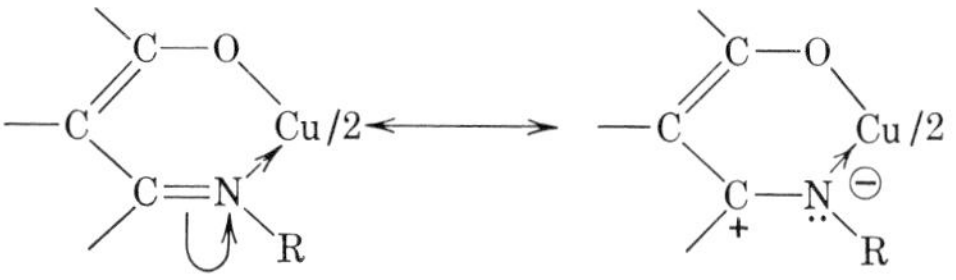

The electron-deficient carbon would be susceptible to attack by amine:

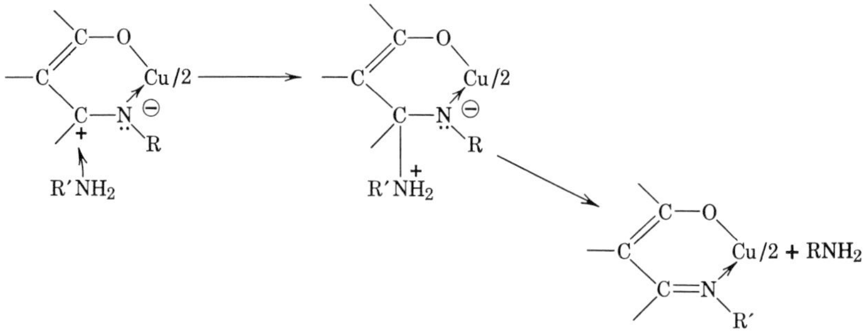

The strength of the metal–nitrogen bond is obviously an important factor: the stronger the bond, the less likely amine exchange becomes. This is consistent with the observation that amine exchange occurs in an ammoniacal solution of $CH_3COCH_2C(=NC_6H_5)CH_3$ when nickel(II) is used, but not when copper(II) ion is present (4). It has been suggested (4) that the amine-exchange process for β-ketoimines involves the reaction of a monochelated species, formed either directly or by dissociation. The driving force for the reaction is probably the formation of the more stable metal-chelate compound; also, the equilibrium would be shifted by loss of volatile amine. Additional discussion of the scope of amine exchange involving β-ketoimines may be found elsewhere (4,29).

D. Chelate Exchange

This method may be defined as the displacement of a chelate from the coordination sphere of a metal ion by another chelate, HCh′:

$$MCh_n + nHCh' \rightleftarrows MCh_n' + nHCh$$

Experimentally, the method consists in heating a mixture of ligand and metal-chelate compound in an appropriate solvent. Frequently, the extent of reaction can be ascertained from the color of the solution. The desired product crystallizes or is caused to precipitate by addition of another solvent. Chelate exchange has been used to prepare β-dicarbonyl–metal compounds (35,36). The extension of this method to the preparation of metal derivatives of salicylaldimines (34), and of β-keto-imines (16,17) shows several advantages: Side reactions, such as amine exchange and hydrolysis, are avoided; a purer sample is likely since contamination by extraneous ions is minimized; the operations are simple; and, it seems to be possible to prepare "mixed" complexes of the type MChCh′ (37).

There are several disadvantages: The chelate-exchange equilibrium may be unfavorable and the desired product unobtainable. The successful isolation of a product requires that the metal-chelate compound be the most insoluble species present. Usually, this requirement is satisfied. Finally, chelate exchange requires the preparation of both a ligand and a metal-chelate compound. Compared with direct reaction, the additional time required for chelate exchange, though not great, may not seem justified.

The time is justified in the preparation of $[C_6H_5COCHC(=NC_6H_5)-CH_3]_2Cu$ because by the direct reaction method, using ammonia as a base, the amine-exchange product, $[C_6H_5COCHC(=NH)CH_3]_2Cu$, is obtained. By contrast, the chelate-exchange process occurs readily and a good yield of product is obtained (16).

The existence of an unfavorable equilibrium is a more serious disadvantage than the amount of time involved. The unfavorable equilibrium might be surmounted by use of an excess of ligand, though this may introduce the problem of separation. Also, if necessary the initial metal-chelate compound can be varied. For example a much higher yield of $[CH_3COCHC(CH_3)=NCH_2CH(CH_3)N=C(CH_3)CHCOCH_3]Ni$ was obtained (37) when the metal-chelate compound used was bis(4-imino-pentane-2-ono)nickel(II) (88%) instead of bis(N-butylsalcylaldimino)-nickel(II) (30%).

The mechanism of chelate exchange has not been elucidated. Moore and Young have measured equilibrium constants for exchanges involving β-diketones and β-dicarbonyl–copper(II) compounds (35). Also, some preliminary kinetic measurements for β-ketoimines and β-ketoimine–copper(II) compounds have been communicated (48).

The preparation of optically active copper-chelate compounds by means of an asymmetric synthesis (38) is suggestive of the importance of chelate exchange to biochemical systems and will be of help in elucidating the mechanisms of chelate exchange. The copper-chelate compound

should exist as three isomers, two enantiomers and a *meso*-form; because the two benzene rings do not lie in the plane of the complex, the structure is nonsymmetrical.   The *levo*-compound was prepared through the reaction of copper-*d*-tartrate and the sexadentate ligand (prepared *in situ*), the *dextro*-rotatory-form was obtained from sodium-bis(*l*-glutamo)copper(II), and the *meso*-compound was not isolated (38).

E. Metal Exchange

In this procedure a metal-chelate compound is mixed with a metal salt in solution, and a more stable metal-chelate compound is formed:

$$\text{MCh}_n + \text{M}'^{n+} \rightleftarrows \text{M}'\text{Ch}_n + \text{M}^{n+}$$

The advantages of the method are the ease of operation and the opportunity to prepare a variety of compounds from one metal-chelate compound.   The method has the disadvantage that the more stable metal-chelate compound is obtained, unless some means of circumventing equilibrium conditions is found.   Furthermore, even if the equilibrium is favorable there is the possibility of contamination of the product by metal ions or by anions.

The possibilities of metal exchange as a synthetic method have not been exploited.   The potential of the method is suggested by the success with which a variety of compounds of the type $\text{Cr}[\text{RCOCHC}(=\text{NR}')\text{R}'']_3$ have been prepared using $\text{Cr}(\text{THF})_3\text{Cl}_3$ and $\text{K}[\text{RCOCHC}(=\text{NR}')\text{R}'']$ (10).   Also, Pfeiffer and co-workers (39) have used the method to determine the relative stabilities of bis-salicylalethylenediiminometal(II) compounds; the order of replacement is $\text{Cu} > \text{Ni} > \text{Zn} > \text{Mg}$.

F. Template Syntheses

A metal-chelate compound or the coordination sphere of a metal ion can serve as a template and induce different ligand molecules to orient in a manner that is suitable for condensation and complex formation. In short, the metal ion serves to organize the reactants in a form that is suitable for compound formation. There are several examples of this, and a few should serve to indicate the potential scope of the method.

The condensation of α-diketones and mercaptoethylamine mainly afforded thiazolines rather than the desired tetradentate ligands, but in the presence of nickel(II) ion, the condensation resulted in good yields of the desired product (40):

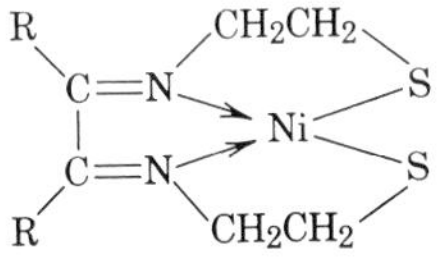

The sulfur atoms in this complex have been alkylated with benzyl bromide with the formation of a monomeric compound; alkylation with α,α'-dibromo-o-xylene produces a macrocyclic compound (41):

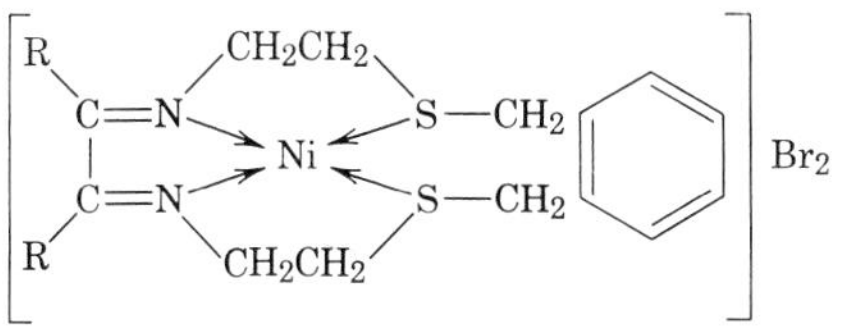

In this example, a coordination compound serves as the template.

Related to this is the formation of compounds of the type Ni[RCO-CR'C(R")=NCH$_2$]$_2$ by amine-catalyzed condensation of β-diketones (or substituted salicyladehydes or o-hydroxyacetophenones) with bis-ethylenediaminenickel(II) chloride (42). The latter is a dimer with chlorine bridges (43).

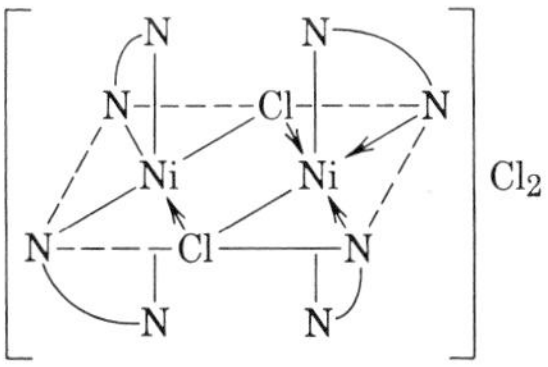

It has been suggested (42) that addition of a catalytic amount of co-ordinating amine such as pyridine (py), but not 2,6-dimethylpyridine, results in bridge-breaking and the formation of [Ni(en)$_2$(py)$_2$]Cl$_2$. An oxygen atom from each of two different β-diketone molecules could then displace the coordinated pyridine molecules and condense with an adjacent amino group. Apparently, bridge-breaking and the formation of a labile compound, $[Ni(en)_2(py)_2]Cl_2$, is necessary because it was observed that β-diketones do not condense with $[Ni(en)_3]Cl_2$ in the absence of suitable amine.

Clearly, a different mechanism must be invoked to explain the condensation of acetone with $[Ni(en)_3](ClO_4)_2$ (44). The condensation products are square-planar diamagnetic nickel(II) compounds having two ethylenediamine molecules per nickel ion and two, three, or four additional isopropylidene (C_3H_4) units.

These examples should serve to suggest the possibilities afforded by template syntheses.

G. Other Methods

Finally, several methods which do not fit into any of the preceding categories deserve mention because of potential usefulness.

1. Interaction of a Soluble Salt of a Schiff Base and a Soluble Salt of a Metal

This method has been used for the syntheses of metal derivatives of β-diketones (23), but it has not been widely used for the preparation of metal derivatives of ketimines and aldimines.

2. Reactions of Functional Groups

The reactions of coordinated ligands represent a useful synthetic method which has not been exploited. The method may involve reactions of functional groups or emplacement of functional groups on the coordination compound.

Transesterification and amidation reactions have been performed on bis(salicylaldimine)copper(II) compounds; transesterification with butyl mercaptan was not successful (29).

The emplacement of functional groups on the coordinated β-ketoimine represents a potentially useful synthetic method. For example, certain β-ketoimine-chromium(III) compounds undergo a facile reaction with N-bromosuccinimide (NBS). It is likely that further examples of this method will be reported.

$(R = C_6H_5, \ p\text{-}CH_3C_6H_4, \ o\text{-}CH_3C_6H_4)$

III. EXPERIMENTAL PROCEDURES

A. Direct Reaction

1. Bis(N,N'-disalicylethylenediimine)-μ-aquodicobalt(II)

The preparation of the ligand (95% yield) and the preparation of the metal-chelate compound (90% yield) is described in *Inorganic Syntheses* (13).

2. Bis(4-iminopentane-2-ono)nickel(II), [CH₃COCHC(=NH)CH₃].Ni

The preparation of the metal-chelate compound by treatment of an ethanolic solution of the ligand with ammoniacal nickel nitrate is described in *Inorganic Syntheses* (16). The procedure is analogous to that used for the preparation of the copper(II) derivative. The synthesis of the nickel(II) compound was reported independently by Archer (18).

B. Constituent Combination

A method which has proven very useful is one developed by Charles

(27,28). This method is applicable to the synthesis of copper, nickel, zinc, and cobalt(II) derivatives of N-alkylsalicylaldimines.

1. Bis[N-(n-butyl)salicylaldimino]copper(II), [o-OC₆H₄CH(=NCH₂CH₂CH₂CH₃)]₂Cu*

To 100 ml. methanol is added successively salicylaldehyde (6.1 g., 0.05 mole), n-butylamine (3.8 g., 0.05 mole), and 50 ml. $1M$ sodium hydroxide. After 15 min., a solution of copper(II) nitrate, 3-hydrate (6.05 g., 0.025 mole) in 100 ml. of water is added, and the solution is stirred vigorously. After cooling in an ice-water bath, the crude product is collected by filtration. Recrystallization from 95% ethanol affords about 8.6 g. of dark-green crystals. Additional product (1.1 g.) is obtained by adding water (about 35 ml.) to the boiling ethanol filtrate (about 70 ml.) to induce crystallization. Total yield 9.7g.(91%), m.p. 79–81 °C.; reported (27): 80.5–81 °C.

2. Bis[N-(n-butyl)salicylaldimino]cobalt(II), [o-OC₆H₄CH(=NCH₂CH₃CH₂CH₃)]₂Co†

To 200 ml. methanol is added successively salicylaldehyde (12.2 g., 0.10 mole), n-butylamine (7.6 g., 0.10 mole), and 100 ml. $1M$ sodium hydroxide. Nitrogen is bubbled through the solution for 15 min. This flask is stoppered and allowed to stand. Meanwhile, nitrogen is bubbled through 200 ml. water for 5 min., and cobalt(II) nitrate, 6-hydrate (14.7 g., 0.05 mole) is dissolved in the water. The resulting solution is added to the salicylaldehyde solution. Nitrogen is bubbled through the solution for 15 min. The mother liquor is removed by means of a filter stick, and the product is recrystallized from acetone using a nitrogen atmosphere. The compound is obtained as red needles (16.2 g., 75%), m.p. 151–153 °C. *Anal.*: Calcd. for $C_{22}H_{28}O_2N_2Co$: C, 64.20; H, 6.81; Found: C, 63.91; H, 6.85.

C. Amine Exchange

The following synthesis was suggested by the method described by Muto (34). The compound was prepared by another method (45), but melting point data were not given.

1. Bis-salicylalethylenediiminocopper (II), (o-OC₆H₄CH=NCH₂)₂Cu

To a solution of recrystallized bis(N-n-butylsalicylaldiimino)copper-(II) is added anhydrous ethylenediamine (0.6 g., 5 mmole). A color

---

* This synthesis was checked by Arthur W. Struss (37).

† This synthesis was developed by Arthur W. Struss (37).

change from dark to light green is noted. Upon warming on a steam bath for 5 min. or less, the product precipitates, is collected by filtration, and is dried *in vacuo*. The yield is 0.7–0.8 g. (85–97%) of light-green platelets, m.p. 315 (dec.). *Anal.* (unrecrystallized material): Calcd. for $C_{16}H_{14}O_2N_2Cu$: C, 58.27; H, 4.28; N, 8.50; Found: C, 57.94; H, 4.42; N, 8.33.

D. Chelate Exchange

1. Bis-(3-Phenylimino-1-phenylbutane-1-ono)copper(II), $[C_6H_5COCHC(=NC_6H_5)CH_3]_2Cu$

The synthesis of this metal-chelate compound (64% yield) is described in *Inorganic Syntheses* (16). The procedure appears to be generally applicable to the syntheses of copper(II)- and nickel(II)-β-ketoimine compounds.

E. Metal Exchange

1. Tris(4-toluidino-3-pentene-2-ono)chromium(III), $[CH_3(4-CH_3C_6H_4NH)C=CHCOCH_3]_3Cr$ (10)

The preparation of this compound is described in *Inorganic Syntheses* (46) and is obtained from interaction of the potassium salt of the β-keto-imine and trichlorotris(tetrahydrofuran)chromium(III). The reaction appears to be generally useful for the preparation of chromium(III)-β-ketoimine compounds.

F. Template Syntheses

1. General Method

Two drops of pyridine are added to a solution of bis-ethylenediamine-nickel(II) chloride (2.49 g. in 20 ml. water) (47). The β-diketone or salicylaldehyde (0.02 mole) can be added directly or as a solution in 20 ml. methanol. The mixture is allowed to reflux for 2 hr., during which time the product crystallizes. The product is collected by filtration: frequently, the material obtained is of analytical purity, and the yield is nearly quantitative.

2. Bis-acetylacetoneethylenediiminonickel(II), $[CH_3COCH(CH_3)C=NCH_2]_2Ni$

Following the general procedure and using 1.98 g. acetylacetone, orange-brown crystals are obtained (82% yield), m.p. 198 °C.; reported (42): 198, 200 °C.

3. *Bis(o-hydroxyacetophenone)ethylenediiminonickel*(II),
*[o-OC₆H₄C(CH₃)=NCH₂]₂Ni** *(42)*

Two drops of pyridine are added to a solution of bis-ethylenediamine-nickel(II) chloride (2.49 g.) in 20 ml. water.   An immediate color change (blue to green) is noted.   Then, *o*-hydroxyacetophenone (2.76 g., 0.02 mole) is added, and the mixture is heated at reflux temperature for 2 hr., during which time the brown crystalline product forms.   The product is collected by filtration, m.p. 290°C. (81% yield).   Analytical data are recorded elsewhere (42).

REFERENCES

1. Martell, A. E., R. L. Belford, and M. Calvin, *J. Inorg. Nucl. Chem.* **5,** 170 (1958).
2. Hovey, R. J., J. J. O'Connell, and A. E. Martell, *J. Am. Chem. Soc.* **81,** 3189 (1959).
3. Gash, V. W., Abstracts of Papers, 138th National Meeting of the American Chemical Society, New York, Sept. 11–16, 1960, p. 58P.
4. Martin, D. F., *Advan. Chem. Ser.* **37,** 192 (1963).
5. Pfeiffer, P., W. Offermann, and H. Werner, *J. Prakt. Chem.* **159,** 313 (1941).
6. Eichhorn, G. L., and J. C. Bailar, Jr., *J. Am. Chem. Soc.* **75,** 2905 (1953).
7. Baker, A. W., and A. T. Schulgin, *J. Am. Chem. Soc.* **81,** 1523 (1959).
8. Das Sarma, B., and J. C. Bailar, Jr., *J. Am. Chem. Soc.* **76,** 4051 (1955).
9. Cromwell, N. H., *Chem. Rev.* **38,** 83 (1946).
10. Collman, J. P., and E. T. Kittleman, *Inorg. Chem.* **1,** 499 (1962).
11. Martin, D. F., G. A. Janusonis, and B. B. Martin, *J. Am. Chem. Soc.* **83,** 73 (1961).
12. Morgan, G. T., and J. D. M. Smith, *J. Chem. Soc.* **1925,** 2030.
13. Diehl, H., and C. C. Hach, in *Inorganic Syntheses,* Vol. 3, L. F. Audrieth, ed., McGraw-Hill, New York, 1950, p. 196.
14. Goodwin, H. A., and F. Lions, *J. Am. Chem. Soc.* **81,** 6415 (1959).
15. Holtzclaw, H. F., Jr., J. P. Collman, and R. M. Alire, *J. Am. Chem. Soc.* **80,** 1100 (1958).
16. Struss, A. W., and D. F. Martin, in *Inorganic Syntheses,* Vol. 8, H. F. Holtzclaw, Jr., ed., McGraw-Hill, New York, to be published.
17. Hseu, T. M., D. F. Martin, and T. Moeller, *Inorg. Chem.* **2,** 587 (1963).
18. Archer, R. D., *Inorg. Chem.* **2,** 292 (1963).
19. Holtzclaw, H. F., Jr., private communication.
20. Holm, R. H., *J. Am. Chem. Soc.* **82,** 5632 (1960).
21. Lions, F., and K. V. Martin, *J. Am. Chem. Soc.* **80,** 3858 (1958).
22. McCarthy, P. J., R. J. Hovey, K. Ueno, and A. E. Martell, *J. Am. Chem. Soc.* **77,** 5820 (1955).
23. Fernelius, W. C., and B. E. Bryant, in *Inorganic Syntheses,* Vol. 5, T. Moeller, ed., McGraw-Hill, New York, 1957, p. 105.
24. Nunez, L. J., and G. L. Eichhorn, *J. Am. Chem. Soc.* **84,** 901 (1962).

* The details of this synthesis are due to Edward J. Olszewski.

25. Martell, A. E., and M. Calvin, *Chemistry of the Metal Chelate Compounds*, Prentice-Hall, Englewood Cliffs, New Jersey, 1953, p. 424.
26. Pfeiffer, P., Th. Hesse, H. Pfitzner, W. Scholl, and H. Thielert, *J. Prakt. Chem.* 149, 219 (1937).
27. Charles, R. G., *J. Org. Chem.* 22, 677 (1957).
28. Charles, R. G., *J. Inorg. Nucl. Chem.* 9, 145 (1959).
29. Verter, H. S., and A. E. Frost, *J. Am. Chem. Soc.* 82, 85 (1960).
30. Figgins, P. E., and D. H. Busch, *J. Am. Chem. Soc.* 82, 820 (1960).
31. Delepine, M., *Bull. Soc. Chim. France* 21, 943 (1899).
32. Schiff, H., *Ann. Chem.* 150, 193 (1869).
33. Pfeiffer, P., E. Buchholz, and O. Bauer, *J. Prakt. Chem.* 129, 163 (1931).
34. Muto, Y., *Nippon Kagaku Zasshi* 76, 252 (1955); through *Chem. Abstr.* 51, 17559 (1957).
35. Moore, T. S., and M. W. Young, *J. Chem. Soc.* 1932, 2694.
36. Wolf, L., E. Butter, and H. Weinelt, *Z. Anorg. Allgem. Chem.* 306, 87 (1960).
37. Struss, A. W., and D. F. Martin, *J. Inorg. Nucl. Chem.* 25, 1409 (1963).
38. Das Sarma, B., and J. C. Bailar, Jr., *J. Am. Chem. Soc.* 77, 5476 (1955).
39. Pfeiffer, P., H. Theilert, and H. Glaser, *J. Prakt. Chem.* 152, 145 (1939).
40. Thompson, M. C., and D. H. Busch, *J. Am. Chem. Soc.* 84, 1762 (1962).
41. Thompson, M. C., and D. H. Busch, Abstracts of Papers, 142nd National Meeting, American Chemical Society, Atlantic City, New Jersey, Sept., 1962, p. 13 N.
42. Olszewski, E. J., L. J. Boucher, R. W. Oehmke, J. C. Bailar, Jr., and D. F. Martin, *Inorg. Chem.* 2, 661 (1963).
43. Antsyshkima, A. S., and M. A. Porki-Koshits, *Dokl. Akad. Nauk., SSSR* 143, 105 (1962); through *Chem. Abstr.* 57, 2954 (1962).
44. Blight, M. M., and N. F. Curtis, *J. Chem. Soc.* 1962, 1204.
45. Pfeiffer, P., E. Breith, E. Lübbe, and T. Tsumaki, *Ann. Chem.* 503, 84 (1933).
46. Collman, J. P., and E. T. Kittleman, in *Inorganic Syntheses*, Vol. 8, H. F. Holtzclaw, Jr., ed., McGraw-Hill, New York, to be published.
47. State, H. M., in *Inorganic Syntheses*, Vol. 6, E. G. Rochow, ed., McGraw-Hill New York, 1960, p. 198.
48. Martin, D. F., *Chem. Ind. (London)* 1963, 1528.

CHAPTER 4

Metal Carbonyls

JACK C. HILEMAN

El Camino College, Torrance, California

CONTENTS

I. INTRODUCTION

This chapter has been written for professional chemists who may wish to prepare metal carbonyls for some phase of their research program, but who have not had extensive previous experience with this class of compound. Thus, some aspects of the procedures have been given in considerable detail when it was considered likely that the reader's prior experience was limited; but in most instances familiarity and skill with routine preparative procedures have been presumed.

Furthermore, it was assumed that the most efficient way to procure metal carbonyls was to purchase them when commercial sources were available. Only preparative procedures for some of those metal carbonyls which cannot be purchased have been described in detail in this chapter. However, references to the original literature sources have been included for the preparation of all metal carbonyls. Some general

TABLE I

The History of Metal Carbonyl Syntheses

Date	Compound	Method	Discoverer	Ref.
1890	$Ni(CO)_4$	Ni plus CO	Mond, Langer, and Quincke	6
1891	$Fe(CO)_5$	Fe plus CO	Mond and Quincke	7
1905	$Fe_2(CO)_9$	Condensed $Fe(CO)_5$	Dewar and Jones	8
1906	$Fe_3(CO)_{12}$	Condensed $Fe_2(CO)_9$	Dewar and Jones	9
1910	$Mo(CO)_6$	Mo plus CO	Mond, Hirtz, and Cowap	10
1910	$Co_2(CO)_8$	Co plus CO	Mond, Hirtz, and Cowap	10
1910	$Co_4(CO)_{12}$	Condensed $Co_2(CO)_8$	Mond, Hirtz, and Cowap	10
1926	$Cr(CO)_6$	Grignard on $CrCl_3$	Job and Cassal	11
1928	$W(CO)_6$	Grignard on WCl_6	Job and Rouvillois	12
1936	$Ru(CO)_5$	RuI_3 plus Ag and CO	Manchot and Manchot	13
1936	$Ru_3(CO)_{12}$[a]	Condensed $Ru(CO)_5$	Manchot and Manchot	13
1940	$Ir_2(CO)_8$	$IrCl_3$ plus Cu and CO	Hieber and Lagally	14
1940	$Ir_4(CO)_{12}$	Condensed $Ir_2(CO)_8$	Hieber and Lagally	14
1941	$Re_2(CO)_{10}$	Re_2O_7 plus CO	Hieber and Fuchs	15
1943	$Rh_2(CO)_8$	Rh plus CO	Hieber and Lagally	16
1943	$Rh_4(CO)_{12}$	Condensed $Rh_2(CO)_8$	Hieber and Lagally	16
1943	$Rh_6(CO)_{16}$[a]	Rh plus Ag and CO	Hieber and Lagally	16
1943	$Os(CO)_5$	OsO_4 plus CO	Hieber and Stallman	17
1943	$Os_3(CO)_{12}$[a]	Condensed $Os(CO)_5$	Hieber and Stallman	17
1954	$Mn_2(CO)_{10}$	MnI_2 plus Mg and CO	Brimm, Lynch, and Sesny	18
1959	$V(CO)_6$	VCl_3 plus Mg–Zn and CO	Natta, Ercoli, Calderazzo, Alberola, Corradini, and Allegra	19
1961	$Tc_2(CO)_{10}$	Tc_2O_7 plus CO	Hileman, Huggins, and Kaesz	20a
1962	$(CO)_4Co-Mn(CO)_5$	$NaCo(CO)_4$ plus $Mn(CO)_5Br$	Joshi and Pauson	20b
1962[b]	$Tc_3(CO)_{12}$[b]	$NaTc(CO)_5$ plus H_3PO_4	Kaesz and Huggins	21a
1963	$(CO)_5Mn-Re(CO)_5$	$Mn(CO)_5$ Br plus $NaRe(CO)_5$	Nesmeyanov, Anisimov, Kolobova, and Kolomnikov	22

[a] The originally reported formulas of $M_2(CO)_9$ and $Rh_4(CO)_{11}$ were found to be $M_3(CO)_{12}$ and $Rh_6(CO)_{16}$, respectively, by Corey and Dahl (23,24).

[b] The rhenium analog was found to be a hydride (21b).

references to survey articles on metal carbonyls have been listed for the interested reader (1–5,5a).

A. Known Metal Carbonyls

The compounds formed between carbon monoxide and zero-valent transition metals have excited the curiosity of chemists ever since Mond discovered nickel carbonyl in 1890. An indication of the persistence with which research has been directed toward the metal carbonyls is given by Table I which lists the known species in the order of their discovery.

The extent to which metal carbonyl syntheses have been investigated can be summarized efficiently by reference to the periodic table. As shown in Table II, many of the transition metals have been incorporated

TABLE II

Transition Metal Families Which Form Carbonyls and
Carbonyl Derivatives

IV-B	V-B	VI-B	VII-B	VIII-B			I-B
Ti	$V(CO)_6$ Bl.–Grn.	$Cr(CO)_6$ Col.	$Fe(CO)_5$ Col. $Mn_2(CO)_{10}$ Yell. $Fe_2(CO)_9$ Yell. $Fe_3(CO)_{12}$ Grn.–Blk.	$Co_2(CO)_8$ Orange $Co_4(CO)_{12}$ Grn.–Blk.	$Ni(CO)_4$ Col.	Cu	
Zr	Nb	$Mo(CO)_6$ Col.	$Tc_2(CO)_{10}$ Col. $Tc_3(CO)_{12}$ Tan	$Ru(CO)_5$ Col. $Ru_3(CO)_{12}$ Yell.	$Rh_2(CO)_8$ Orange $Rh_4(CO)_{12}$ Red $Rh_6(CO)_{16}$ Blk.	Pd	Ag
Hf	Ta	$W(CO)_6$ Col.	$Re_2(CO)_{10}$ Col.	$Os(CO)_5$ Col. $Os_3(CO)_{12}$ Yell.	$Ir_2(CO)_8$ Yell.–Grn. $Ir_4(CO)_{12}$ Yell.	Pt	Au

into metal carbonyls. Also, compounds possessing the carbon monoxide group along with other ligands have been prepared from certain transition metals, even though it was not possible to prepare the "parent" carbonyl which contained nothing but the metal and carbon monoxide. Such metals have been indicated by shading the appropriate section of Table II.

B. Metal Carbonyl Structures

Investigation of the nature of the chemical binding between a neutral molecule like carbon monoxide and a zero-valent metal has been a large part of the research on metal carbonyls. All metal carbonyls discovered up to now have the general formula $M_x(CO)_y$ and can be placed in one of nine groups according to their molecular formula as shown in Figure 1. In addition, the internal bonding of these molecules has been shown to have even fewer possible variations and has been considered to possess three types of linkage: (1) covalent metal-to-metal bonds with each metal supplying one electron; (2) coordinate covalent, σ-type metal-to-carbon monoxide bonds, with the carbon atom donating a pair of electrons to the metal and receiving a partial π-type "back-bonding" from

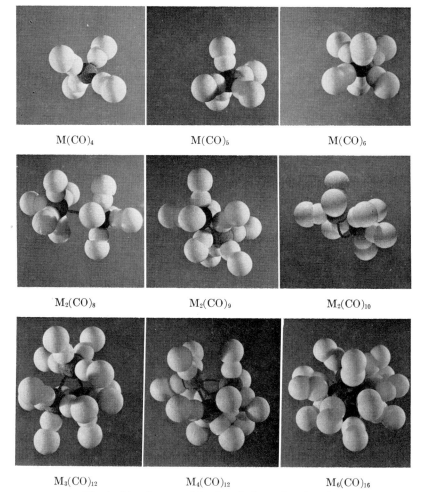

$$M(CO)_4 \qquad M(CO)_5 \qquad M(CO)_6$$

$$M_2(CO)_8 \qquad M_2(CO)_9 \qquad M_2(CO)_{10}$$

$$M_3(CO)_{12} \qquad M_4(CO)_{12} \qquad M_6(CO)_{16}$$

Fig. 1. Styrofoam models of typical metal carbonyls.

the d electrons of the metal; and (3) covalent, σ-type metal-to-carbon monoxide bonds with the carbon atom acting as a bridge between two metal atoms and forming a bond with each. There is recent x-ray (24) evidence that carbonyls of the type $M_6(CO)_{16}$ possess internal bonding of a fourth and still different type, possibly involving metal–monoxide bonds in which the carbon atom serves as a bridge between three metal atoms. Also, it should be noted that the tetragonal pyramid of $W(CO)_5$ has not been considered in this discussion because of its short half-life (2 min.) and because, as yet, it has not been isolated from solution (25).

Most of the metal carbonyls have demonstrated the utility of the Sidgwick rule regarding the effective atomic number (E.A.N.) of the transition metals involved. In essence, the E.A.N. rule suggests that each metal should react in a manner to acquire the same number of external electrons as the subsequent inert gas in the Periodic Table. For example, nickel reacts with four carbon monoxide groups to acquire an interest in eight electrons beyond nickel's original 28, making a total of 36 in correspondence with the atomic number of krypton. Metals with uneven atomic numbers form carbonyls which must utilize metal-to-metal bonds or bridging carbonyls or else display the paramagnetic effects associated with unpaired electrons. During the early history of the research on metal carbonyls, all of them were found to obey the E.A.N. rule and, thus, to be diamagnetic. However, the recent discovery of vanadium carbonyl (19), illustrated the possibility of non-adherence to the E.A.N. rule, and the resulting paramagnetism confirmed the presence of an unpaired electron. Apparently, $Tc_3(CO)_{12}$ also "violates" the E.A.N. rule, but the magnetic properties have not been investigated thoroughly (21). The most recently discovered $M_6(CO)_{16}$ compounds represent such a large departure from the E.A.N. rule that one hesitates to speculate on the magnetic implications (24).

C. Predicted but Undiscovered Metal Carbonyls

The most obvious group of potential new metal carbonyls would seem to be based on those metals which are known to form metal–carbon monoxide bonds, but which also contain other ligands. These were indicated in Table II by the shaded areas. There exist compounds such as $[Pd(CO)Cl_2]_n$ and $[Pt(CO)Cl_2]_n$ in Group VIII-B, $Cu(CO)X$, Ag_2SO_4-(CO), and $Au(CO)Cl$ in Group I-B, $(C_5H_5)_2Ti(CO)_2$ in Group III-B, and $[Na(C_6H_{14}O_3)_2][Ta(CO)_6]$ and $[Na(C_6H_{14}O_3)_2][Nb(CO)_6]$ in Group V-B. The fact that such metal–carbonyl bonds can be stabilized by the presence of the other ligands, suggests strongly that experimental con-

ditions exist under which the parent carbonyls may be stable. Podall
(26) indicated that the thermal stability of carbonyls decreased in going
from left to right across the Periodic Table according to the following
order:

$$\text{Even } Z: \quad Cr(CO)_6 > Fe(CO)_5 > Ni(CO)_4$$

$$\text{Odd } Z: \quad Mn_2(CO)_{10} > Co_2(CO)_8$$

Also, the ease of formation of the metal–carbonyl bond, based on the
rate of reaction of metal with carbon monoxide, was suggested to be

$$\text{Even } Z \text{ (across):} \quad Cr < Fe < Ni$$

$$\text{Even } Z \text{ (down):} \quad Cr < Mo < W$$

$$\text{Odd } Z \text{ (across):} \quad V < Mn < Co$$

It was noted that a carbonyl of copper, if one could be made, would
have the formula $Cu_2(CO)_6$ and would contain either a Cu–Cu bond or a
combination of both Cu–Cu and bridging carbonyl bonds. The existence of a carbonyl of zinc with a formula $Zn(CO)_3$ was considered to be
unlikely in view of the stable configuration of the $3d^{10}$ electrons and lack
of opportunity for bond hybridization into a p^3 configuration of CO
ligands around the zinc atom. In fact, the formation of a $Zn(CO)_4{}^{++}$
cation was considered more probable than the formation of the parent
carbonyl.

Group III-B and IV-B metals were considered more likely to form
metal–carbonyl anions that were more stable than the parent carbonyls,
because of the greater back-coordination of d electrons to the CO ligands.
Ormont (27) suggested, from thermodynamic considerations, that
metals of Group II-B could form tricarbonyls, but carbonyls of Sn, Ge,
Mg, Ca, Sr, and Ba would be unstable. Pospekhov (28) developed a
formula, extending the E.A.N. rule, which successfully predicted the
formula of the then known polymeric metal carbonyls, and suggested
the possibility of preparing $Ag_2(CO)_6$ and $Cu_2(CO)_6$.

Another source of new carbonyls is inherent in the fact that several of
those presently known belong to a "homologous" series. For example,
the three iron carbonyls, $Fe(CO)_5$, $Fe_2(CO)_9$, and $Fe_3(CO)_{12}$ represent
increasingly complex members of a series derivable from the simplest
member. It seems quite likely that still higher members of the series
exist, perhaps not easily separable from the lower members by sublimation or other conventional separation techniques. Metlin, Wender, and
Sternberg (29a) found CO absorption by $Co_2(CO)_8$ at high pressures;
but the resulting species was not stable enough to recover. Hieber
(29b) has reported the existence of the anion $Fe_4(CO)_{13}{}^{--}$ for which the

corresponding metal carbonyl, $Fe_4(CO)_{14}$, has never been found. Such high molecular weight members would tend to be insoluble in the usual solvents and thus be unavailable for routine identification by infrared analyses. As in the case of the early history of organic polymers, perhaps we inorganic chemists should investigate the "gunk" that lies in the bottom of our reaction vessels.

Ruthenium and osmium also formed a series of carbonyl compounds, and for many years the existence of $Ru_2(CO)_9$ and $Os_2(CO)_9$ was taken for granted, in analogy to $Fe_2(CO)_9$. However, careful x-ray studies by Corey and Dahl (23) resulted in the assignment of the formulas $Ru_3(CO)_{12}$ and $Os_3(CO)_{12}$ to the compounds previously reported as nonacarbonyls. The question immediately arises, where are the ruthenium and osmium congeners of $Fe_2(CO)_9$? If routine synthetic procedures convert the $Ru(CO)_5$ and $Os(CO)_5$ directly to the dodecacarbonyls, will more sophisticated procedures permit the isolation of the now seemingly hypothetical nonacarbonyls?

Joshi and Pauson (20b) took an imaginative approach to metal carbonyl synthesis when they prepared a dimeric carbonyl containing mixed metals, $(CO)_4Co–Mn(CO)_5$. It was obvious that they only scratched the surface of the possibilities; and soon after the announcement of the Co-Mn compound, Nesmeyanov et al. (22) announced the synthesis of the lemon-yellow Mn–Re species, $(CO)_5Mn–Re(CO)_5$. There seems to be no reason for limiting this approach to dimers nor to dimetallic carbonyls.

Finally, the lanthanide and actinide elements surely deserve consideration as a source of potential metal carbonyls. The only serious effort reported in this direction was an unsuccessful attempt to prepare a uranium carbonyl (30a). It is difficult to predict the ability of underlying "f" orbitals to become involved in back-donation π-bonding or their effect on the few available electrons in overlying d orbitals, but some of the ingredients for metal–carbonyl bonding seem to be present.

In the 1960 Tilden Lecture, Nyholm (30b) evaluated the factors pertinent to metal–ligand bond stability, and his comments have application to the problem of "discovering" new metal carbonyls. It was suggested that "(a) the metal atom must have available electrons in suitable d-orbitals. Hence for the nonbonding configuration d^n the number of *possible* double bonds decreases as n approaches zero (the nonexistence of carbonyls of the nontransition can thus be explained). (b) The ligand must have empty d- (as in R_3P) or p-orbitals which can be made available (as in CO) to receive these d-electrons. (c) The sizes of the orbitals must be such as to ensure effective overlap." Nyholm further stated that the ability to form π-d bonds varies thusly:

84   J. C. HILEMAN

Fig. 2. Externally stirred high-pressure autoclaves. **a.** Autoclave Engineers, Inc., 300 cc. autoclave; **b.** Parr Instrument Co., Series 4500 pressure reaction apparatus.

$$Ni^0 \gg Pt^0 > Pd^0$$

$$Au^+ > Cu^+ > Ag^+$$

Zero-valent metals > Bivalent > Tervalent

A possible explanation for the failure of $Pt(CO)_4$ or $Pd(CO)_4$ to be prepared was given in terms of heats of atomization, promotion energies to the spin-paired state, and the relative ease with which an electron can be removed from the d^{10} nonbonding shell.

II. EQUIPMENT REQUIREMENTS

A. High Pressure

In general, the equipment required for high pressure carbonylation reactions is similar to that used for hydrogenation. In alphabetical order, the three main U.S.A. sources of equipment are: American Instrument Company, Inc., 8030 Georgia Avenue, Silver Springs, Maryland; Autoclave Engineers, Inc., 2903 West 22nd Street, Erie, Pennsylvania; and Parr Instrument Company, 211 Fifty-third Street, Moline, Illinois. Each offers an extensive line of well-designed instruments and accessories; and a few examples, of utility in metal carbonyl syntheses, will be briefly described in terms of one man's experience, with the full realization that alternate procedures and equipment may be equally efficacious.

Typical externally stirred autoclaves are shown in Figure 2. The advantages associated with externally stirred pressure reactors reside in the flexibility of accessories available. Sampling probes can be kept in operation while stirring proceeds, auxiliary internal cooling coils may give protection against "runaway" reactions, glass liners can be used to constrain liquid solutions, and considerable flexibility regarding the bomb capacity is possible. The only significant limitation is imposed by the maximum pressure that can be tolerated, usually listed as being from 1000 to 6000 p.s.i.g., depending on the temperature and on the manufacturer. Because carbon monoxide cylinders are delivered with a maximum pressure of 68 atm. (1000 p.s.i.g.), it is possible to perform many experiments using the tank as the only source of pressure. However, if the full 1000-p.s.i.g. tank pressure is used to fill an autoclave at room temperature and then the autoclave in heated, pressures beyond the capacity of the equipment may be developed. In such cases, external stirring might not be possible and it would be necessary to utilize an accessory bomb head without a stirring gland. Thus, some externally stirred autoclaves may not permit the user to take full advantage of the cheapest available source of pressure, namely, the original carbon monoxide tank.

a

b

Fig. 3. Rocking-type high-pressure autoclaves. **a.** Parr Instrument Co., Series 4000 Rocker Apparatus, 500 or 1000 ml. capacity; **b.** American Instrument Co., Micro Series, high-pressure shaking assembly, 43–400 ml. capacity.

a

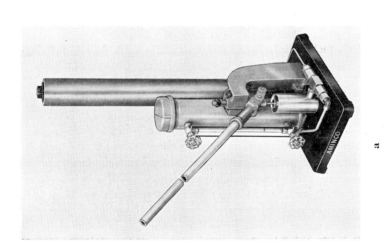

b

Fig. 4. High-pressure pumps. **a.** American Instrument Co., hand-operated gas booster pump; **b.** American Instrument Co., motor-driven gas booster pump.

To utilize higher pressures, autoclaves are designed in such a way that the contents are mixed by a rocking (shaking) or by a rotating (rolling) motion.   Examples of rocking equipment are shown in Figure 3.   The high-pressure rotating autoclave which Hieber used for much of his work has been described in the literature (30c).   Depending on the temperature, apparatus of this type can usually withstand pressures considerably beyond those conventionally employed in metal carbonyl syntheses.   In fact, one cannot utilize a very large portion of the pressure capacity of these autoclaves without auxilliary pumping equipment to "boost" the initial carbon monoxide pressure available from the tank. Hand pumps are available for pressurizing small vessels, and motor-driven hydraulic pumps are useful for autoclaves with volumes of 1–2 l. Examples of such pumps are shown in Figure 4.   Lightweight dia-phragm-typecompressors are available also for developing pressures up to 30,000 p.s.i.g.

A detailed discussion of the rocker-type autoclave and the accessory booster pump has been presented by Jolly (31) in a laboratory book that also discusses many related aspects of the problems encountered in high-pressure syntheses.

There have been no literature reports of metal carbonyl syntheses at pressures in the region of 100,000 p.s.i.g.   However, attempts to syn-thesize some of the more "obstinate" carbonyls may well benefit from such high pressures.   At such conditions, the volumes available for the reactions become limited, but equipment is commercially available which can pressurize a 100 ml. space to 100,000 p.s.i.g.

Equipment of the "opposed-anvil" type, Figure 5, has been made available recently, capable of being operated at pressures above one million p.s.i.g.; but there is no record of its use for metal carbonyl syntheses.   Presumably, only solids could be handled, but the possi-bility of the high-pressure conversion of one solid carbonyl to another is intriguing.   There is a severe limitation on the volume of the sample that can be pressurized, the maximum being about 0.5 cm.[3].

A few comments regarding the use of high-pressure autoclaves may be of value to chemists without previous experience.   With many rocking-type shaker autoclaves, the seal is achieved by screwing a cap down until it squashes a copper gasket between the cap and the body of the bomb. It is important to anneal the copper gasket so that it will deform into the space which it should fill.   A convenient way to anneal the copper, while keeping it free of a brittle oxide coating, is to heat the gasket red hot in a Fisher burner flame and drop it into a beaker containing sufficient acetone to cover the gasket.   The reducing action of the acetone keeps the oxide coating to a minimum.   Such an annealed gasket can be used

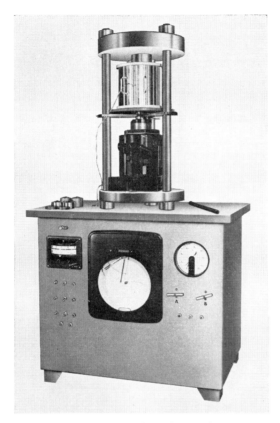

Fig. 5. Tem-Pres Research Inc., opposed-anvil press for pressures to 2.5 million p.s.i.g.

many times if it is not flattened too much by excessive pressure from the cap screws. Other metals such as silver and stainless steel are sometimes used as gaskets.

Several safety considerations deserve attention. Silver rupture disks are usually installed in the system to protect gauges and the carbon monoxide tank itself, especially if an auxilliary pump is used to elevate the pressure above that of the full tank. Such rupture disks should be positioned so as to blow away from the operator in case of an explosion. Because of the compressibility of carbon monoxide (or any gas), failures in the equipment are much more explosive than is the case with hydraulic liquids.

Many autoclaves are designed so that the operator can turn on the motors and heaters, as well as control the temperatures and pressures, without being in the inclosure that contains the autoclave. A movable

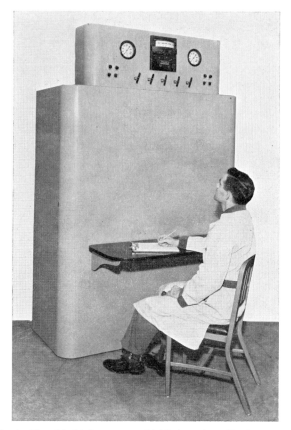

Fig. 6. Movable high-pressure shield, Autoclave Engineers, Inc.

safety enclosure, especially designed by Autoclave Engineers, Inc. for use with high pressure autoclaves is shown in Figure 6.

Resist the temptation to immerse the steel bomb-case in liquid nitrogen in order to condense a reactant into the bomb. It is believed that such low temperatures can cause the steel to "crystallize" with the result that the strength is permanently decreased, even after the bomb-case is warmed to room temperature. As a final safety note, it should be remembered that carbon monoxide is both extremely poisonous and flammable. The *Matheson Gas Data Book* (32) describes the precautions to be taken when using carbon monoxide.

B. Auxiliary Equipment for Metal Carbonyl Syntheses

Many of the operations involved in metal carbonyl syntheses are facilitated by the use of vacuum-line techniques, but it is not within the scope of this chapter to go into detail about the routine use of such techniques. However, a few specialized procedures may be worth mention.

Because of the high vapor pressure of carbon monoxide at the temperature of liquid nitrogen, about 400 mm. Hg at −196°C., it is not possible to quantitatively transfer carbon monoxide in a vacuum line without a specially designed pump. Most frequently, a Toepler pump is used, along with an automatic electronic control system. The previously mentioned book by Jolly describes such a system (31). The automatic Sprengler pump developed by Bartocha et al. (33) has also been used

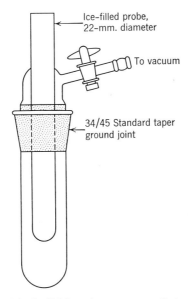

Fig. 7. Sublimation apparatus (34).

for noncondensable gas transfer. The ability to pump quantitative volumes of noncondensable gases permits one to determine the per cent carbon monoxide in a metal carbonyl which is thermally decomposed into the metal and carbon monoxide.

Fractional sublimation at reduced pressures is widely used as a technique for separating and purifying metal carbonyls. An excellent, though inexpensive and simple, sublimation apparatus has been described by King and Stone (34), and is diagramed in Figure 7.

A "dry-box" is extremely helpful in many operations related to the preparation of air-sensitive reagents used in metal carbonyl syntheses, and in some cases the carbonyl itself requires careful handling away

92   J. C. HILEMAN

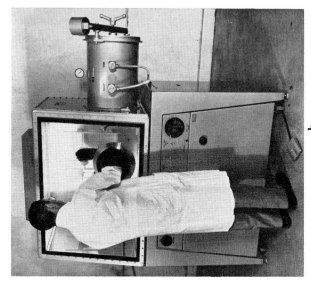

b

a

Fig. 8. Inert atmosphere "glove-boxes." **a.** Manostat Corp. Plexiglass Model 41-905; **b.** D. L. Herring Corp. "Inert Atmosphere Laboratory" Model HE-93.

from air. The term "dry-box" covers a wide range of experimental arrangements, varying considerably in effectiveness and cost. At one extreme has been the use of inexpensive, nitrogen-filled, polyethylene bags whose contents were manipulated from outside the bag. Low-cost modifications of this approach have been described by Shore (35) and by Ocone and Simkin (36). A more elaborate plexiglass glove-box is commercially available in the medium price range, and is shown in Figure 8a. The most effective types of "dry-boxes" are more accurately described as "inert atmosphere laboratories." The example shown in Figure 8b continuously recirculates an inert gas over heated copper and through a molecular sieve, keeping the oxygen concentration below 1 part per million (p.p.m.) and the water-vapor volume below 0.1 p.p.m. If necessary, balances and other equipment can be installed inside of the "dry-box."

Occasionally it is desirable to perform a reaction in an atmosphere of high-pressure carbon monoxide but without letting the other reactants come in contact with the metal bomb walls. For some situations, glass liners supplied by the autoclave equipment manufacturers are satisfactory, but in other cases a modified Grosse flask is useful (37). The Grosse flask consists of a test tube equipped with a capillary spiral having a cross section so small that the mean free path of high-pressure reactants will not permit diffusion out of the reaction portion of the tube during the experiment. Figure 9 shows the Grosse flask used in a 1-l. bomb

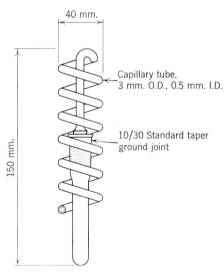

Fig. Modified Grosse flask for high pressure carbon monoxide reactions (39).

to prepare $Tc_2(CO)_{10}$ from TcO_2 and to prepare halide derivatives of that carbonyl (38).   Simpler forms of the Grosse tube with a straight piece of capillary tubing containing a single "U" bend have also been effective.

An apparatus for determining the molecular weights of metal carbonyls has been adapted to use with the high-vacuum line by Kobayashi et al. (40a).   The vapor-pressure change of a solvent containing a dissolved substance is compared against the pure solvent.   The heart of the apparatus is shown in Figure 10,

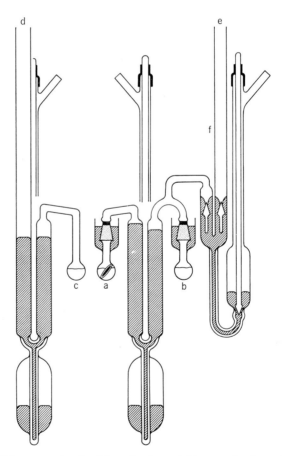

Fig. 10. High vacuum tensimetric molecular weight apparatus for metal carbonyls. **a**. Sample bulb and stirrer; **b**. pure solvent bulb; **c**. tensimeter, solvent V.P.; **d**. to manometer; **e**. to high vacuum; **f**. stock-type Hg seal.

and an additional constant temperature bath and cathetometer are needed to complete the system. Cryoscopic methods have been known to give errors amounting to nearly 100% in the case of some metal carbonyls, and with those same troublesome compounds this tensimetric method resulted in accuracies within 10%, using extremely small quantities, *in vacuo*.

III. PROCEDURES FOR THE PREPARATION OF METAL CARBONYLS

A. Literature References to Metal Carbonyl Preparations

Many syntheses for metal carbonyls have been reported which will not be specifically referred to in the subsequent procedures described in this chapter, especially those involved in patents or in manufacturing commercially available compounds. The methods used to discover each metal carbonyl have been indicated earlier in Table I, but these were occasionally quickly superseded by more effective processes. Sometimes the method described in the research literature has been dictated by considerations of available equipment, a desire to investigate the extent of utility of a particular procedure, or "vested interest" in one of the ingredients used as a reactant. Therefore, it may be that an investigator will wish to pursue the problem of metal carbonyl syntheses further than is possible from the procedures described in this chapter. To this end, Table III has been compiled as a summary of the research literature on metal carbonyl synthetic methods, good, bad, or indifferent, which have come to the attention of the author. The procedures summarized in Table I have been repeated, for convenience. Table III includes references to articles in books which specialize in synthetic methods, such as *Inorganic Syntheses*, because some of these books are more generally accessible than the journals with the original research papers.

B. Preparing Iron Carbonyls

The commercial availability of iron pentacarbonyl makes it relatively easy to prepare diiron nonacarbonyl and triiron dodecarbonyl, even without high-pressure carbonylation. For many subsequent syntheses, the higher molecular weight species have been considered superior to pentacarbonyl. However, very little information is available on the thermodynamic properties of the diiron and triiron carbonyls, most of the data having been obtained with $Fe(CO)_5$. Some of the mechanistic studies involving $Fe(CO)_5$ have implications for the higher carbonyls.

TABLE III

Literature References to Metal Carbonyl Syntheses

Metal carbonyl	Method	Ref.
$Co_2(CO)_8$	Co + CO	10
	$CoCO_3$ + H_2 + CO	31,40b
	$Co(CO)_4H$ + Heat	41
	CoO + CO	42
	$Co(NH_3)_6Cl_2$ + CO	43
	Co_2O_3 + Synthesis gas	44
	Cobalt salts + Cu(Ag) + CO	30c
	$HCo(CO)_4$ + Olefin + CO	45
$Co_4(CO)_{12}$	$Co_2(CO)_8$ + Heat	10,30c,46–48
$Cr(CO)_6$	$CrCl_3$ + Grignard + CO	11,41,49–56
	$CrCl_3$ + Et_3Al + CO	57,58a,58b
	$CrCl_3$ + Na benzophenone ketyl + CO	59
	$CrCl_3$ + Mg + I_2 + CO	18,60
	Chromium(III) acetylacetonate + Mg + CO	61,61b
	$CrCl_3$ + Al + Al_2Cl_6 + CO	62,63
	$Na_2[Cr_3(CO)_{14}]$ + CO	64
	$CrCl_3$ + Na + CO	65–68,69a,69b
	$CrCl_3$ + $LiAlH_4$ + CO	70
$Fe(CO)_5$	Fe + CO	7,8
	$FeSO_4$ + NH_3 + CO	43
	$Fe(CO)_4I_2$ + Cu	41,71a
$Fe_2(CO)_9$	Fe(CO) + Light	8,41,47,71b
$Fe_3(CO)_{12}$	$Fe_2(CO)_9$ + Heat	9
	$HFe(CO)_4^-$ + "Active" MnO_2	41,71c
	$HFe(CO)_4^-$ + Oxidizing agents	72,73
	$H_2Fe_3(CO)_{11} \cdot NR_3$ + Acid	74,75
$Ir_2(CO)_{18}$	$IrCl_3$ + Cu + CO	14
$Ir_4(CO)_{12}$	$Ir_2(CO)_8$ + Heat	14
$Mn_2(CO)_{10}$	MnI_2 + Mg(Cu) + CO	18,76
	$Mn(C_2H_3O_2)_2$ + Na benzophenone ketyl + CO	77
	$Mn(C_2H_3O_2)_2$ + Et_3Al + CO	57
	$MnCl_2$ + PhenylMgX + CO	78
	$MnCl_2$ + Na + CO	79
	π–$CH_3C_5H_5Mn(CO)_3$ + Na + CO	80a,80b
	$Mn(CO)_5NO_3$ + Heat	81
$Mo(CO)_6$	Mo + CO	10
	$MoCl_5$ + Grignard + CO	41
	$MoCl_5$ + Et_3Al + CO	57,58a,58b
	$MoCl_5$ + Na + CO	65,69b
	$MoCl_5$ + Zn + CO	54,83
	$MoCl_5$ + Al + CO	84
	$MoCl_5$ + $Fe(CO)_5$	85
	$Cr(CO)_6$ + U_3O_8 fission products	86
	$Na_2 Mo_3(CO)_{14}$ + CO	64
	$Na_2 Mo_2(CO)_{10}$ + CO	87

(continued)

TABLE III (*continued*)

Metal carbonyl	Method	Ref.
$Ni(CO)_4$	$Ni + CO$	6,88,89
	$Ni(NH_3)_6Cl_2 + CO$	43
	$H_2Ni_2(CO)_6 + Acid$	90
	$H_2Ni_2(CO)_6 + Base$	91
	$(C_3H_8NiBr)_2 + CH_3OH + CO$	92
	$NiSe + base + CO$	93
	NiS or $Ni(CN)_2 + CO$	94,95
	$Ni(\text{II})$ salts $+$ Reducing agents $+ CO$	41,96–98, 99a,99b
$Os(CO)_5$	$OsO_4 + CO$	17
$Os_3(CO)_{12}$	$Os(CO)_5 + Heat$	17,23,100
$Re_2(CO)_{10}$	$Re_2O_7 + CO$	15,38
	$Re_2S_7 + CO$	15
	$KReO_4 + CO$	15
	$HRe(CO)_5 + Heat$	101
$Rh_2(CO)_8$	$Rh + CO$	16
$Rh_4(CO)_{12}$	$Rh_2(CO)_8 + Heat$	16
$Rh_6(CO)_{16}$	$RhCl_3 + Cu(Ag) + CO$	16,24
$Ru(CO)_5$	$RuI_3 + Ag + CO$	13
$Ru_3(CO)_{12}$	$Ru(CO)_5 + Heat$	13,23
$Tc_2(CO)_{10}$	$Tc_2O_7 + CO$	20
	$TcO_2 + CO$	38
$Tc_3(CO)_{12}$	$NaTc(CO)_5 + H_3PO_4$	21
$V(CO)_6$	$VCl_3 + Mg\text{–}Zn + CO$	19,102
	$Mg[V(CO)_6]I_2 + HCl$	103
	$[Na(diglyme)_2][V(CO)_6] + H_3PO_4$	104
	Ditoluene vanadium $+ CO$	105
$W(CO)_6$	$WCl_6 + Grignard$	12,41,54
	$WCl_6 + Zn + CO$	83
	$WCl_6 + Et_3Al + CO$	57,58a,58b
	$WCl_6 + Fe(CO)_5$	106
	$WCl_6 + FeCO_5 + H_2$	107
	Tungsten(v) halide $+ Na + CO$	65,69b

Table IV summarizes the data on the properties of these carbonyls.

The reaction of $Fe(CO)_5$ to form $Fe_2(CO)_9$ is believed to involve the preliminary formation of a highly active $Fe(CO)_4$ species. Although never isolated from solution, infrared spectral evidence for $Fe(CO)_4$ has been obtained by Stolz et al. (25) in u.v. irradiated solutions of $Fe(CO)_5$.

$$Fe(CO)_5 + Ultraviolet \rightarrow Fe(CO)_4 + CO$$

Presumably $Fe(CO)_4$ could also result from the decomposition of iron carbonyl hydride.

$$Fe(CO)_4H_2 \rightarrow Fe(CO)_4 + H_2$$

TABLE IV

Properties of the Iron Carbonyls

Property	$Fe(CO)_5$	$Fe_2(CO)_9$	$Fe_3(CO)_{12}$
Color	Colorless[a]	Yellow	Black
Mol. Wt.	196	364	504
M.P., °C.	−20	—	—
B.P., °C.	103	d.	d.
Sublimation, °C., and Pressure	—	—	60,0.1mm.
Decomposition, °C.	130	100	140
Density, g./cc.	1.52	2.08	2.00
M–M bond length, A.	—	2.46	—
M–C bond length, A.	1.82	1.9	—
C–O bond length, A.	1.14	1.15	—
		1.3	
$\Delta H°$ (Formation) kcal.	−182.6	—	—
$\Delta H°$ (Decomposition) kcal.	138.3[b]	—	—
$\Delta H°$ (Combustion) kcal.	−385.9	—	—
$\Delta H°$ (Vaporization) kcal.	8.9	—	—
Redox Potentials:			

$$3Fe(CO)_4{}^{--} = Fe_3(CO)_{12} + 6e^- \qquad E = -0.74 \text{ v.}$$

[a] Most sources state that $Fe(CO)_5$ is straw-to-yellow colored, apparently due to traces of yellow $Fe_2(CO)_9$.

[b] For the reaction: $Fe(CO)_5(g) = Fe(g) + 5CO(g)$.

The subsequent reactions of $Fe(CO)_4$ with $Fe(CO)_5$ has been considered to be the source of $Fe_2(CO)_9$.

$$Fe(CO)_4 + Fe(CO)_5 \rightarrow Fe_2(CO)_9$$

However, the isotopic carbon monoxide exchange studies of Keeley and Johnson (47) suggested that the photodissociation process was not a major factor in the reaction path; and an alternative mechanism involving an activated $Fe(CO)_5{}^*$ molecule was proposed.

$$Fe(CO)_5 + \text{Ultraviolet} \rightarrow Fe(CO)_5{}^*$$

$$Fe(CO)_5{}^* + Fe(CO)_5 \rightarrow Fe_2(CO)_9 + CO$$

The mechanism by which $Fe_3(CO)_{12}$ is formed has not been elucidated, and the process involved when it is formed by heating $Fe_2(CO)_9$ may be different than when it is formed in aqueous solutions by oxidation of $Fe(CO)_4H_2$ or $Fe(CO)_4{}^{--}$. The thermal decomposition of $Fe_2(CO)_9$ in organic solvents at 95°C. may involve the same activated intermediates discussed in the reactions of $Fe(CO)_5$, with the net result considered to be the overall reaction (1):

$$3Fe_2(CO)_9 \rightarrow Fe_3(CO)_{12} + 3Fe(CO)_5$$

In aqueous solutions, the iron carbonylate anions have been found to be interconvertible depending on pH and temperature. Above pH 10, $Fe_2(CO)_8^{--}$ tends to be stable, and in the pH range of 5–10, $Fe_3(CO)_{11}^{--}$ is obtained.

A procedure for the preparation of $Fe_2(CO)_9$ has been described (47) in which a sample of $Fe(CO)_5$ was transfered to an evacuated pyrex tube and sealed off. The bottom (liquid) portion of the tube was kept dark, while the top (gaseous) portion was exposed to sunlight for 2 hr. Solid Fe_2-$(CO)_9$ accumulated on the illuminated walls of the tube until all of the liquid $Fe(CO)_5$ had disappeared, resulting in essentially a quantitative yield. Because of the simplicity of this syntheses and the availability of other, equally effective and simple procedures (Table III), no further attention to the methods of preparing $Fe_2(CO)_9$ was considered necessary in this chapter. However, the reader is cautioned about the poisonous nature of $Fe(CO)_5$ and the necessity of allowing sufficient space in the pyrex tube to accommodate the pressure buildup of the released carbon monoxide.

The preparation of $Fe_3(CO)_{12}$ also avoids the necessity of high-pressure carbonylation; and the well-executed procedures of King and Stone in *Inorganic Syntheses* (see Table III) have been summarized to emphasize the techniques involved in the preparation of an especially active form of manganese dioxide oxidizing agent and the storage of an air-sensitive metal carbonyl. The reaction was said to proceed as indicated in these equations (which have no significance in terms of mechanism):

$$Fe(CO)_5 + 3OH^- \rightarrow HFe(CO)_4^- + CO_3^{--} + H_2O$$

$$3HFe(CO)_4^- + 3MnO_2 + 3H_2O \rightarrow Fe_3(CO)_{12} + 3Mn^{++} + 9OH^-$$

Using conventional safety procedures involving a hood, safety pan, and (presumably) a shield, a 2-l. flask was equipped with a stirrer, nitrogen inlet, and reflux condenser. To the nitrogen-flushed flask was added 60 g. (42 ml.) $Fe(CO)_5$ (commercial grade), 170 ml. methanol, and 135 g. of an aqueous solution containing 45 g. sodium hydroxide. The exothermic reaction mixture was stirred for 30 min., after which 125 ml. saturated ammonium chloride solution was added.

Commercial MnO_2 was not a suitable oxidizing agent for this buffered $HFe(CO)_4^-$ solution, so a special form of MnO_2 was prepared by cautiously treating a solution of 67 g. $KMnO_4$ in 300 ml. water with 100 ml. of 95% ethanol. Stirring sometimes initiated a vigorous reaction; and, if not, the mixture was heated gently to start the process. The purple $KMnO_4$ solution was converted to a brown paste of MnO_2.

The MnO_2 paste was added to the flask, stirred, and allowed to react.

with the evolution of heat, for about 2 hr.   To the flask was added a solution made up of 250 ml. of $1M$ sulfuric acid in which was dissolved 40 g. $Fe_2SO_4 \cdot 7H_2O$.   The iron(II) salt reduced the excess MnO_2.   To the resulting red solution, 300 ml. of $9M$ sulfuric acid was added; and the mixture was stirred until the $Fe_3(CO)_{12}$ precipitated completely, leaving a pale-green solution.   The filtered product was washed with hot $1M$ sulfuric acid, followed by 95% ethanol, and then a low-boiling paraffin hydrocarbon solvent such as pentane.   As much as 45 g. of product resulted (90% yield, based on iron pentacarbonyl), sufficiently pure for most purposes.   To avoid air oxidation, the $Fe_3(CO)_{12}$ was stored under nitrogen, and stability was improved by leaving the last traces of solvent on the solid.   Highly pure $Fe_3(CO)_{12}$ was obtained as a black sublimate at 60°C. and 0.1 mm. Hg pressure or by pentane extraction under nitrogen in a Soxhlet extraction apparatus.   When heated much above 60°C., the $Fe_3(CO)_{12}$ decomposed to the metal.   Extended storage of the highly purified product has been known to result in the formation of pyrophoric decomposition products.

Alternate procedures based on the acidification of ammine complexes of $H_2Fe_3(CO)_{11}$ have also given excellent yields of $Fe_3(CO)_{12}$ (see Table III).

C. Preparing Cobalt Carbonyls

Although not commercially available, the fact that $Co_2(CO)_8$ can be synthesized without high-pressure carbonylation allows one to prepare it, as well as $Co_4(CO)_{12}$, with relative ease.   However, improved yields result, on the basis of the cobalt content of the reactants, when $Co_2(CO)_8$ is prepared by using pressurized mixtures of hydrogen and carbon monoxide.   Therefore the method of Wender et al. (40b), as modified by Jolly (31), will be described in this chapter as an introduction to high-pressure techniques.   Other methods, not requiring high pressure, have been referred to in Table III.

$Co_2(CO)_8$ has been prepared by direct high-pressure carbonylation of the metal and the mechanism is assumed to involve processes very similar to those encountered in the reactions to form iron and nickel carbonyl. However, Podall has suggested that the dimetallic carbonyls, with their bridging carbonyls or metal-to-metal bonds require separate consideration.   The thermodynamic stability of the metal carbonyl bonds may be of less importance than the rate factors influencing the formation of the first metal–carbon monoxide linkage.   The carbonyl exchange studies of Keeley and Johnson (47) do not offer much help in the elucidation of the reaction mechanism because all of the CO groups change immeasurably fast in both light and dark.   Entirely different mechanisms must be involved in the reductive carbonylation of aqueous cobalt(II) ion to $HCo(CO)_4$ and the subsequent conversion of the hydrides to $Co_2(CO)_8$ by heat; but little work has been reported to clarify the details. The physical and thermodynamic data available in the literature are summarized in Table V.

TABLE V
Properties of Cobalt Carbonyls

Property	$Co_2(CO)_8$	$Co_4(CO)_{12}$
Color	Orange	Green-black
Mol. Wt.	342	572
M.P., °C.	51	d.
B.P., °C.	d.	d.
Sublimation, °C. and Pressure	45, 10 mm.	—
Decomposition, °C.	52	60
Density, g./cc.	1.73	—
M–M bond length, A.	2.54	2.50
ΔH (Formation) kcal.	—	33,000[a]
Redox potentials:		
$2Co(CO)_4{}^- = Co_2(CO)_8 + 2e^-$	$E = -0.4$ v.	

[a] For the formation from $Co_2(CO)_8$ rather than from the elements (48).

A typical high-pressure synthesis of $Co_2(CO)_8$ has been based on the reaction:

$$2CoCO_3 + 2H_2 + 8CO \rightarrow Co_2(CO)_8 + 2H_2O + 2CO_2$$

A 300 ml. stainless steel, rocking-type, autoclave was charged with 7.5 g. $CoCO_3$ and 75 ml. of low-boiling paraffin hydrocarbon solvent, flushed of air with carbon monoxide, and pressurized to 3500 p.s.i.g. with a 50–50 mixture of carbon monoxide and hydrogen. The mixture of gases was obtained by first pressurizing to 1700 p.s.i.g. with carbon monoxide, followed by the addition of high-pressure hydrogen until the desired pressure of 3500 p.s.i.g. was reached. The rocking autoclave was heated to 150–160°C., kept there for 3 hr., cooled to room temperature, and vented into a hood. The resulting dark solution was filtered through paper and refrigerated overnight, presumably at about −10°C. Dark red crystals were separated from the solvent by decantation and dried in a stream of dry nitrogen. The yield was about 6 g., accounting for approximately 55% of the original cobalt.

The preparation of $Co_4(CO)_{12}$ is quantitative and routine, involving nothing more than heating $Co_2(CO)_8$ under nitrogen to about 50°C. None of the mechanistic features of the reaction have been clarified, and in fact, the structure is not established beyond possibility of doubt due to the reliance on x-ray data obtained from disordered crystals (107).

$$2Co_2(CO)_8 + Heat \rightarrow Co_4(CO)_{12} + 4CO$$

D. Preparing Technetium Carbonyl

Because of the lack of pertinent thermodynamic data, the procedure used to prepare $Tc_2(CO)_{10}$ (20a) was developed in close analogy to the method by which Hieber and Fuchs prepared $Re_2(CO)_{10}$ (15). The procedure was complicated by the fact that the only technetium isotope available in significant amounts was Tc-99, a weak beta emitter (0.29 M.e.v.) with a half-life of 2.12×10^5 yr. Although such a weak beta emitter was adequately shielded from the research personnel by the walls of any container it was in, it was sufficiently "active" so that constant surveillance of the laboratory with a special thin-window Geiger counter was necessary. Any items of equipment that became contaminated and all refuse associated with the synthesis were disposed of by burial at sea.

Tc-99 can be obtained in several forms from Oak Ridge National Laboratories. The present synthesis used a saturated ammonium pertechnetate solution as a starting point, and evaporation of the water from the colorless solution resulted in white $NH_4^{99}TcO_4$.

A 0.85 g. sample of dry ammonium pertechnetate was weighed into a combustion boat and pushed back to the sealed end of a horizontal Vycor tube, 25 mm. in diameter and about 30 cm. long. The Vycor tube was supported so that a tube furnace could be slid sideways in a manner that would allow the combustion boat to protrude into the furnace. The unheated end of the Vycor tube was connected to a 1-atm. nitrogen source through a dry ice–acetone trap. The Vycor tube was flushed with nitrogen and the cold furnace was allowed to heat rapidly to about $400°C$., pyrolyzing the sample to black TcO_2. Gases formed in the decomposition were evacuated to 1 mm., while being pulled through the CO_2–acetone trap where water and blue N_2O_3 condensed. Most of the TcO_2 formed on the hot walls of the Vycor tube or remained in the combustion boat. After cleaning the CO_2–acetone trap, the Vycor tube was connected to a 1-atm. source of oxygen and the furnace was raised to $600°C$. A yellow Tc_2O_7 formed, along with a trace of orange material (assumed to be $HTcO_4$), and both compounds tended to migrate out of the furnace and condense on the cooler walls of the Vycor tube just outside the furnace. After letting the system cool, the Vycor tube was held vertically and the band of yellow Tc_2O_7 was melted with a glass-blowing torch until it ran back down to the end of the tube, ready for transfer to the autoclave.

A 350 ml., copper-plated steel bomb was flushed with dry nitrogen. With the nitrogen flow continuing, the end of the Vycor tube was placed into the mouth of the bomb and broken with a hammer, allowing the broken glass, combustion boat, and adherent Tc_2O_7 to drop into the bomb. The bomb was purged at 800 p.s.i.g. with Matheson C.P. grade carbon monoxide and pressurized to 3000 p.s.i.g. at room temperature. Without rocking, the bomb was heated to 220°C. for 20 hr., during which time the pressure was 5100 p.s.i.g. After the autoclave cooled back to room temperature, the carbon monoxide was vented to the roof through a CO_2–acetone trap, designed to keep Tc-99 compounds from escaping.

The autoclave interior was washed with three successive 50 ml. portions of diethyl ether which dissolved the $Tc_2(CO)_{10}$. The broken glass and combustion boat and a black residue were transferred to a $6M$ nitric acid solution for recovery of Tc-99 waste, and the bomb interior was also quickly rinsed with dilute nitric acid. The bomb interior retained measurable radioactivity.

The ether was removed from the solution of $Tc_2(CO)_{10}$ at reduced pressure, using a manostated water aspirator. Previous experience with ether solutions of $Cr(CO)_6$, $Mn_2(CO)_{10}$, and $Re_2(CO)_{10}$ suggested that such a solvent stripping process tended to co-distil the metal carbonyl, unless the ether vapor was passed over a 0°C. internal condenser of the type shown in Figure 11. The ether residue was sublimed at 50°C. and 10^{-2} mm. of Hg in the apparatus shown in Figure 7. The yield was about 0.10 g. of colorless crystals, representing approximately 10% of the original Tc-99. The $Tc_2(CO)_{10}$ melted at 159–160°C. in a sealed capillary.

An alternate procedure (38) was developed using the TcO_2 recovered from the nitric acid solutions of residues resulting from the Tc_2O_7 process. The nitric acid was replaced with sulfuric acid by careful evaporation to near dryness, care being taken not to loose volatile $HTcO_4$. The resulting mixture was neutralized to a phenolphthalein end point with $1M$ NaOH, excess acetaldehyde was added, and the mixture was kept warm for several hours until flocculent TcO_2 precipitated. The TcO_2 was filtered, dried at room temperature, and carbonylated in a Grosse flask (Fig. 9).

In a typical experiment, 0.50 g. of TcO_2 was pressurized to 300 p.s.i.g. at room temperature and heated to 250°C., without rocking, for 24 hr. About 0.50 g. of $Tc_2(CO)_{10}$ was recovered from the bomb, representing a yield of nearly 50%, based on the metal.

E. Preparing Manganese Carbonyl

The procedures reported for the syntheses of $Mn_2(CO)_{10}$ represent a marked departure from those used thus far in this chapter. A subtle interaction of reducing agents, coordinating solvents, and high-pressure carbonylation has been required to give useful yields, and Podall coined the phrase "reductive carbonylation" to describe the general process. "Reductive carbonylation" is effective in the production of several other metal carbonyls such as $V(CO)_6$ and $Cr(CO)_6$, and the application of the technique to the synthesis of $Mn_2(CO)_{10}$ will demonstrate its features.

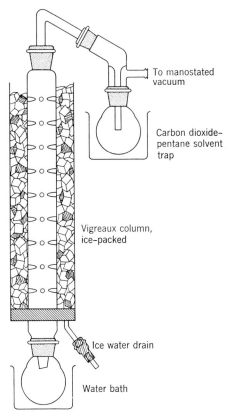

To manostated vacuum

Carbon dioxide-pentane solvent trap

Vigreaux column, ice-packed

Ice water drain

Water bath

Fig. 11. Reduced-pressure solvent stripping still.

The original Grignard-type of synthesis of $Mn_2(CO)_{10}$ by Brimm, Lynch and Sesny (18), and the Friedel–Craft type of syntheses of Cr-$(CO)_6$, by Fischer and Hafner (Table III) can be construed as variations on the "reductive carbonylation," and in all cases the goal seems to be that of obtaining an organo-metallic intermediate with a metal in a low oxidation state and with ligands and coordinated solvent molecules which can be displaced by high-pressure carbon monoxide. In the case of $Mn_2(CO)_{10}$, good yields have been obtained with widely divergent approaches (Table III), one of which, the triethylaluminum reduction method of Podall, Dunn, and Shapiro, will be described in detail in this chapter.

The physical and thermodynamic properties of $Mn_2(CO)_{10}$ have been collected in Table VI.

TABLE VI
Properties of $Mn_2(CO)_{10}$

Property	$Mn_2(CO)_{10}$
Color	Yellow
Mol. Wt.	470
M.P., °C.	151
B.P., °C.	d.
Sublimation, °C., and Pressure	50, 10^{-2} mm.
Density, g./cc.	1.81
M–M bond length, A.	2.93
$\Delta H_{(Formation)}$, kcal.	−400.9
Redox Potentials:	
$2Mn(CO)_5{}^- = Mn_2(CO)_{10} + 2e^-$ $E = -0.68$ v.	

A reaction mechanism for the formation of metal carbonyls by the alkyl aluminum route has been proposed (57) and can be illustrated as follows:

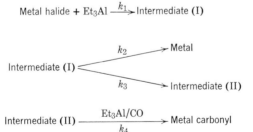

Metal halide + Et_3Al $\xrightarrow{k_1}$ Intermediate (I)

Intermediate (I) $\xrightarrow{k_2}$ Metal
Intermediate (I) $\xrightarrow{k_3}$ Intermediate (II)

Intermediate (II) $\xrightarrow[k_4]{Et_3Al/CO}$ Metal carbonyl

It was suggested that the relative rate constants were in the order $k_3 > k_2 > k_1$ which meant that the production of intermediate(I), assumed to be an ethyl-metal compound, was the rate-determining step. The formation of intermediate(II) was thought to be rapid, as indicated by the small amount of free metal formed. The step involving k_4 was considered to be more heavily influenced by steric effects than the step involving k_1; and, in fact, k_4 may have consisted of a sequence of several other steps.

The mechanism of Grignard syntheses of metal carbonyls has been reviewed by several chemists (1–3,51) without final agreement. There was consensus about the formation of a hydrocarbon–metal intermediate, prior to hydrolysis; and all workers agreed, at least in the case of Cr-$(CO)_6$, that the metal carbonyl was formed only after hydrolysis of the Grignard reaction mixture.

In the preparation of $Mn_2(CO)_{10}$, manganese(II) acetate tetrahydrate, 49 g. (0.20 mole) was dried by azeotropic distillation with 100 ml. benzene in an apparatus designed to allow the water to be separated from the refluxing benzene. The anhydrous $Mn(OAc)_2$ became lumpy unless stirred with a mechanical stirring blade during the distillation. An attempt to use a magnetic stirrer was unsuccessful. The lumpy Mn-$(OAc)_2$ could be crushed easily with a mortar and pestle while still wet with benzene. After filtering, the last traces of benzene were removed in a vacuum desiccator at about 0.01 mm. Hg, and the dried product was stored in the desiccator over Drierite.

Diisopropyl ether was dried by vacuum distillation from lithium aluminum hydride, and 150 ml. was used to prepare a suspension of the anhydrous $Mn(OAc)_2$. The dry ether was also used to prepare a separate solution of triethylaluminum (T.E.A.) containing 120 ml. (0.80 mole) of T.E.A. (Ethyl Corporation, commercial grade) in 300 ml. of diisopropyl ether.

The addition of T.E.A. to ether was exothermic and was performed at 0°C., under an atmosphere of nitrogen. Extreme caution was exercised because of the pyrophoric character of T.E.A. Tygon tubing was used to connect the original metal T.E.A. container with the flask of diisopropyl ether, and the addition was made quite slowly. [*Note:* Consideration should be given to the use of triisobutylaluminum (T.I.B.A.) as a replacement for T.E.A. The originators of this synthesis found that T.I.B.A. gave somewhat better yields, and its decreased tendency to be pyrophoric would make it safer to handle.]

The $Mn(OAc)_2$ suspension was placed in a 1-l. three-necked round-bottom flask equipped with a stirrer, a nitrogen source, a connection to the flask of T.E.A. ether solution, and an 0°C. ice bath.

The T.E.A. solution was added slowly to the stirred $Mn(OAc)_2$ suspension over a period of about 2 hr. The resulting dark-brown mixture was transferred under a vigorous flow of nitrogen to a 2 l. copper-plated high-pressure bomb which also had been swept out with dry nitrogen. [*Note:* The originators of this synthesis transferred the $Mn(OAc)_2$–T.E.A.–ether mixture to the bomb in a nitrogen-filled dry-box.] The bomb was purged at 800 p.s.i.g. and pressurized to 3000 p.s.i.g. (room temperature) with carbon monoxide. While being rocked, the bomb was heated to 100°C. over a period of about 1 hr. The originators of this synthesis suggested that a cooling source was used to keep their 1-l. bomb cooled to 80–85°C. until an initial, "violent" reaction occurred between the T.E.A. and $Mn(OAc)_2$. However, no evidence of such a reaction was noted by the present author when heating the autoclave continuously to 100°C. and maintaining that temperature for 6 hr. Perhaps the larger, 2-l. bomb acted as a heat sink to minimize the effect of a sudden energy surge.

After cooling the bomb to room temperature, excess carbon monoxide was vented; and the contents were transferred under nitrogen to a 2-l. three-necked round-bottom flask equipped with a nitrogen inlet through a CO_2 cold-finger trap, a stirrer, and a 500 ml. pressure-equalizing dropping funnel. With the flask packed in ice, the mixture was "quenched" by first slowly adding 400 ml. ice-water and then 100 ml. concentrated hydrochloric acid. The flask was warmed in a water bath and the contents were stripped free of diisopropylether at reduced pressure, using the ice-packed solvent-stripping column shown in Figure 12. $Mn_2(CO)_{10}$ was removed from the remaining aqueous suspension by steam-distillation. The steam condensate was filtered; and the $Mn_2(CO)_{10}$ was warmed to 40°C. and sublimed onto a 0°C. probe at 0.01 mm. Hg, using the sublimation apparatus shown in Figure 8. The original authors reported a 54% yield (21.0 g.), nearly the same as that achieved in the procedure described here. The well-formed golden-yellow crystals were scraped off of the sublimation probe and stored under nitrogen in a screw-capped jar. The melting point was 153–155°C. (uncorrected).

F. Preparing Vanadium Carbonyl

The solvent used in the preparation of a carbonyl has a significant effect on the yields. For example, Natta et al. (61) found that the yields of $Cr(CO)_6$ could be increased above 80% by using pyridine as the complexing-type solvent. This was in keeping with the general concept of "reductive carbonylation" discussed in the Sec. F preparation of $Mn_2(CO)_6$. The Italian chemists have exploited the use of pyridine-type solvents in metal carbonyl syntheses and, using this approach, were the first to synthesize $V(CO)_6$ (19). The present procedure has been designed to utilize the nitrogen-base solvent approach and is closely adapted from the synthesis of $V(CO)_6$ by Ercoli et al. (102).

An oscillating-type, stainless-steel, 500 ml. autoclave was charged with 9.9 g. anhydrous VCl_3, 220 g. dry pyridine, and 16.0 g. activated metal mixture. The activated metal mixture contained 2 g. iodine (as the activating agent) and 14 g. of a 1 to 2.7 mixture of magnesium and zinc powders (3.8 g. Mg and 10.2 g. Zn, respectively). The reactants in the autoclave were stirred vigorously while being pressurized to 135 atm. (about 2000 p.s.i.g.) with carbon monoxide. The autoclave was heated to 135°C., kept there for 8 hr., cooled, and vented. Over the 8-hr. heating period, the pressure dropped from 208 atm. (about 3050 p.s.i.g.) to 160 atm. (about 2350 p.s.i.g.).

The reaction residue was washed from the bomb with pyridine and allowed to stand a day, protected at all times by an atmosphere of nitrogen. The supernatant solution was vacuum distilled at about 1 mm. Hg without heat, and the resulting solid mass was mixed with 200 ml. water and 400 ml. of "very pure" diethyl ether. The resulting mixture was cooled, and acidified with 300 ml. 4M hydrochloric acid with intermittent shaking. (Caution: Hydrogen was being evolved.) The ether layer was quickly separated from the aqueous phase, washed with additional dilute hydrochloric acid followed with water, and dried overnight with $MgSO_4$. The ether solution was vacuum distilled to give a concentrated brown solution which evolved hydrogen. On continued removal of the ether, a crystalline residue resulted which was sublimed at 40–50°C. under 15 mm. Hg to give moist $V(CO)_6$. The $V(CO)_6$ was dried under nitrogen with P_2O_5 and resublimed, care being taken to avoid losses by volatilization or by undue heating. The blue-green product weighed 5.3 g., representing a 38% yield based on the vanadium metal. The $V(CO)_6$ decomposed before melting in a nitrogen-filled, sealed melting point tube, the decomposition range being 60–70°C. Air-sensitive $V(CO)_6$ should be stored under nitrogen.

V. IDENTIFICATION OF METAL CARBONYLS

A. Infrared Spectra

A convenient method of identifying metal carbonyls utilized infrared spectral data. The carbon–oxygen stretching mode, characteristic of the carbonyl group, is sufficiently affected by the kind of metal present and the symmetry properties of the metal carbonyl molecule to permit rather easy identification of most compounds. The pattern, wave number, and intensity of adsorption maxima often correlate well with theoretical concepts of molecular structure, and an extensive literature has been accumulated relative to the infrared spectra of metal carbonyls (108). In general, all metal carbonyls show at least one adsorption maximum in the 2000 cm.$^{-1}$ region, as a reflection of metal–CO interactions. If bridging carbonyls are present, stretching frequencies in the 1800 cm.$^{-1}$ region are invariably found; but the converse is not always true because of the possibility of various types of resonance interactions.

Precise location of maxima varies somewhat with the type of solvent used in obtaining the spectra, the sharpest patterns being obtained with non-polar hydrocarbons; and the general "picture" established by the spectrum of a particular metal carbonyl depends on the optical system of the instrument being used. Therefore, it is advisable to use comparable solvents and infrared optics if it is desired to correlate experimental spectra with those published in the literature.

As an indication of how sensitive the infrared spectral discrimination can be, Figure 12 shows typical patterns of the Group VII-B metal carbonyl congeners, using a double-beamed instrument with lithium fluoride optics and carbon tetrachloride as the solvent.

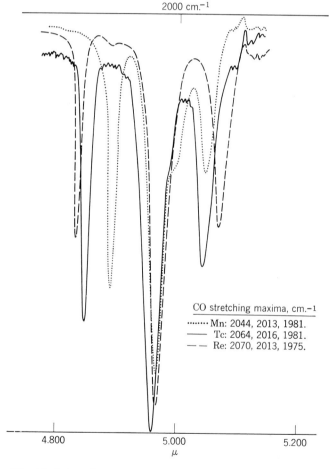

Fig. 12. Infrared spectras of Group VII-B metal carbonyls (5).

Table VII presents some of the literature values of the infrared adsorption maxima of the carbonyl groups in metal carbonyls, listing as many different solvent and optical systems as possible. Some of the older data contained inaccuracies, but have been listed because no more recent values were available with comparable solvents or optics.

B. X-Ray Spectrographic Analyses

The definitive story on the structural identification of a metal carbonyl has invariably involved a careful x-ray study of a single crystal, with the concomitant elaborate analysis of the data, often with the aid of computers. In some cases, severe experimental problems have resulted from the volatility of the metal carbonyl or from the tendency of the x-rays themselves to decompose the sample. X-ray powder pictures have been used occasionally to compare the general patterns of known carbonyls with a new one suspected to have the same crystal structure. [For example, in the assignment of octahedral symmetry to $V(CO)_6$ in analogy with the $Cr(CO)_6$ (127).] Table VIII tabulates the available data on x-ray spectroscopic parameters.

C. Miscellaneous Analytical Techniques

The magnetic properties of a metal carbonyl may be of interest, especially in the case of a newly discovered compound. The nuclear magnetic resonance (NMR) peaks of a hydrocarbon will be shifted in a predictable manner if a paramagnetic species is dissolved in the hydrocarbon; and the coaxial tube method devised by Evans (137) has been effective with metal carbonyls (38). However, the use of a Gouy balance may be more effective in some cases, especially if the hydrocarbon solubility of the metal carbonyl is limited.

Molecular weight measurements have been of fundamental significance in metal carbonyl syntheses. The cryoscopic methods have lead to major discrepancies in some cases (127), but generally gave good results if a suitable solvent was used. A tensimetric vapor pressure method has been described which gave good results with $Tc_2(CO)_{10}$ (38), using the apparatus shown in Figure 10. The newest approach involves the use of differential thermal measurements of the pure solvent as compared with metal carbonyl–solvent mixtures, using equipment such as the Mechrolab vapor pressure osmometer.

Elemental analyses are usually performed by commercial analysts, in the interest of efficiency, but routine CO_2 analytical trains can be used to determine the carbon (and, therefore, carbon monoxide) content. The metals can be determined by standard quantitative procedures, if so

TABLE VII

Infrared Absorption Maxima of the Carbonyl Groups in Metal Carbonyls

Compound	Stretching frequency maxima, cm.$^{-1}$ (s = strong, m = moderate, w = weak, vw = very weak)				Optical system	Solvent or state	Ref.
$Co_2(CO)_8$	2070(s)	2025(s)	1858(m)		NaCl	n-Hexane	109
	2068.8(m)	2058.7(w)	2042.4(s)	2030.7(m)	LiF	n-Heptane	110
	2022.7(m)	2001.7(vw)	1991(vw)	1866.0(w)			
	1857.2(w)						
	2071(m)	2069(m)	2059(s)	2044(s)	Grating	Hexane	111
	2031(m)	2022(m)	1866(w)	1857.5(w)			
	2071	2042	2005	1845	Grating	Chloroform	112
	2072(m)	2069(m)	2059(w)	2042(s)	Grating	CCl_4	111
	2030(m)	2022(m)	1865(w)	1851(w)			
$Co_4(CO)_{12}$	2063.2(s)	2054.5(s)	2037.9(vw)	2027.5(vw)	LiF	n-Heptane	110
	1867.0(w)						
	2103(vw)	2063(s)	2055(w)	2037.5(w)	Grating	Hexane	111
	2027(w)	2018(vw)	1990(vw)				
$Cr(CO)_6$	1987.5(s)	1955(vw)			Grating	Hexane	111
	2122(vw)	2089(vw)	2020(vw)	2000(s)	LiF	Gaseous	113
	1967.5(w)						
	1983.5(s)	1953(vw)			Grating	$C_2H_2Cl_4$	111
	2000.1				LiF	Gaseous	114
	1984				?	Cyclohexane	115
$Fe(CO)_5$	2084(m)	2028(s)	1994(m)	1935(w)	NaCl	Gaseous	116
	2022.9(m)	2000.3(s)	1963.6(w)		LiF	n-Heptane	110
	2034.4	2013.5			LiF	Gaseous	114
	2022(m)	2000(s)	1964(vw)		Grating	Hexane	111

(continued)

J. C. HILEMAN

TABLE VII (*continued*)

Compound	Stretching frequency maxima, $cm.^{-1}$ (s = strong, m = moderate, w = weak, vw = very weak)	Optical system	Solvent or state	Ref.
$Fe(CO)_5$	2034, 2014	?	?	117,118
	2020(m), 1997(s), 1961(vw)	Grating	$C_2H_2Cl_4$	111
	2019, 1998	Grating	Chloroform	112,115
	2021, 2000	?	Cyclohexane	115
$Fe_2(CO)_9$	2087, 2023, 1831	?	?	1
	2080(m), 2034(s), 1828(s)	NaCl	Solid	116
$Fe_3(CO)_{12}$	2046(s), 2023(m)	Grating	Hexane	111
	2047(s), 2024(m), 1865(vw), 1834(vw)	Grating	$C_2H_2Cl_4$	111
	2043(s), 2020(s), 1833(m)	NaCl	Toluene	119
	2043(s), 2020(s), 1997(m), 1858(vw)	LiF	Carbon disulfide	120
	1826(vw)			121
$Mn_2(CO)_{10}$	2033, 2028, 1997	?	?	117
	2074, 2015, 1972	?	?	
	2044(m), 2013(s), 2002(w), 1983(m)	LiF	Cyclohexane	38
	2044(m), 2012(s), 1918(m)	LiF	CCl_4	20
	2043, 2013, 1982	Grating	Cyclohexane	112
	2043, 2010, 1980	Grating	Chloroform	112
$Mo(CO)_6$	1989.5(s), 1957(vw)	Grating	Hexane	111
	2002.6	LiF	Gaseous	114
	1985.5(s), 1955(vw)	Grating	$C_2H_2Cl_4$	111
	2115(w), 2021(m), 1983(s)	LiF	Chloroform	122
	2110(w), 1984, 1983(s)	?	Cyclohexane	115
	2109(vw), 2084(vw), 2020(vw), 2020(vw), 2004(s), 1971(w)	LiF	Gaseous	113,122

Ni(CO)$_4$	2060				NaCl	Gaseous	123
	2046				?	Cyclohexane	115
	2057				?	?	117,118
	2043(s)	2004(vw)			Grating	C$_2$H$_2$Cl$_4$	111
	2045.7(s)	2007.0(vw)			LiF	n-Heptane	110
	2057.6				LiF	Gaseous	114
	2057(s)	2122(vw)	1990(vw)	2018(vw)	LiF	Gaseous	124
Re$_2$(CO)$_{10}$	2049	2013	1983		?	?	117
	2070(m)	2014(s)	2003(w)	1976(m)	LiF	Cyclohexane	38
	2070(m)	2012(m)	1981(m)		LiF	CCl$_4$	20
Tc$_2$(CO)$_{10}$	2064(m)	2017(s)	2005(w)	1984(m)	LiF	Cyclohexane	38
	2064(m)	2016(s)	1981(m)		LiF	CCl$_4$	20
	2064.3(m)	2016.8(s)	1984.2(m)		LiF	Cyclohexane	21
	2065(m)	2018(s)	1982(m)		NaCl	CCl$_4$	125
Tc$_3$(CO)$_{12}$	2093.3(m)	2036.5(s)	2014.4(s)	1989.6(s)	LiF	Cyclohexane	21
V(CO)$_6$	1976				?	Cyclohexane	126
W(CO)$_6$	1997.5				LiF	Gaseous	114
	2101(vw)	2080(vw)	1998(s)	1965(w)	LiF	Gaseous	113

TABLE VIII. Crystal Structure Data for Metal Carbonyls

Compound	Crystal system	Space group	Molecules per cell	Unit cell parameters, Å	Bond distances, Å			Ref.
					M–M	M–C	C–O	
$Co_2(CO)_8$	—	—	—	—	2.54	—	—	107
$Co_4(CO)_{12}$	Orthorhombic	$Pccn$	4	$a = 11.66$; $b = 8.94$; $c = 17.14$	2.50	—	—	107
$CrCO_6$	Orthorhombic	Pna	4	$a = 11.72$; $b = 6.27$; $c = 10.89$	—	1.80	1.15	129
$Fe(CO)_5$	Monoclinic	cc	4	$a = 11.71$; $b = 6.80$; $c = 9.28$; beta $= 107.6°$	—	1.84	—	130,131
$Fe_2(CO)_9$	Hexagonal	$C6_3/m$	2	$a = 6.45$; $c = 15.98$	2.46	1.9 1.8	1.15 1.3	130
$Fe_3(CO)_{12}$	Monoclinic	$P2_1/n$	4	$a = 8.88$; $b = 11.33$; $c = 17.14$; beta $= 97°9.5'$	2.75 2.85	—	—	132
$Mn_2(CO)_{10}$	Monoclinic	Ia or $I2/a$	4	$a = 14.16$; $b = 7.11$; $c = 14.67$; beta $= 105°$	2.92	—	—	133a,133b
$Mo(CO)_6$	Orthorhombic	$Pn2_1a$	4	$a = 12.02$; $b = 6.48$; $c = 11.23$	—	2.13	1.15	129,127
$Ni(CO)_4$	Cubic	$Pa3$	8	$a = 10.84$	—	1.84	1.15	134
$Os_3(CO)_{12}$	Monoclinic	$P2_1/n$	4	$a = 8.10$; $b = 14.79$; $c = 14.64$; beta $= 100°27'$	2.88	1.95	1.14	135
$Re_2(CO)_{10}$	Monoclinic	Ia or $I2/a$	4	$a = 14.70$; $b = 7.15$; $c = 14.91$; beta $= 106°$	3.02	—	—	133a
$Rh_6(CO)_{16}$	Monoclinic	$I2/a$	4	$a = 17.00$; $b = 9.78$; $c = 17.53$; beta $= 121°45'$	2.78	1.86 2.17	1.16 1.20	24
$Ru_3(CO)_{12}$	Monoclinic	—	—	$a:b:c = 0.550:1.00:0.986$; beta $= 100°46'$	—	—	—	23
$Tc_2(CO)_{10}$	Monoclinic	Ia or $I2/a$	4	$a = 14.72$; $b = 7.20$; $c = 14.90$; beta $= 104^1/_2°$	3.02	—	—	38,136
$V(CO)_6$	Orthorhombic	$Pn2_1a$ or $Pnma$	4	$a = 11.97$; $b = 11.28$; $c = 6.47$	—	—	—	127,19
$W(CO)_6$	Orthorhombic	Pna	4	$a = 11.90$; $b = 6.42$; $c = 11.27$	—	2.3	—	129

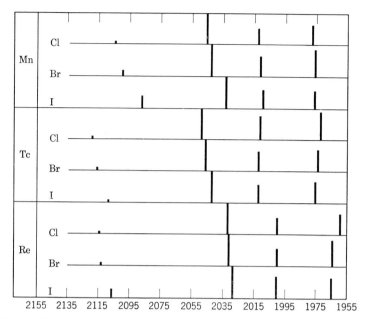

Fig. 13. Position and relative intensities of infrared stretching bands of Group VII-B $M_2(CO)_8X_2$ species. CCl_4 solvent with LiF optics.

desired. The earlier discussion of the Toepler pump illustrated an alternate technique for establishing the per cent metal and carbon monoxide.

Analyses are often predicated on the preparation of metal carbonyl derivatives, perhaps in conjunction with infrared spectral studies. For example, technetium carbonyl was considered likely to be very much like $Mn_2(CO)_{10}$ and $Re_2(CO)_{10}$ in its structure and its reactions with halogens. Infrared spectra of the halide derivatives showed sufficient similarities, Figure 13, to strongly suggest that the formula was indeed $Tc_2(CO)_{10}$.

Some recent work by Lundquist and Cais (138) on the ultraviolet adsorption spectra and extinction coefficients of $Fe(CO)_5$, $Cr(CO)_6$, and $Mn_2(CO)_{10}$ will be useful from an analytical point of view. Earlier u.v. studies had been reported, but most of them were primarily qualitative.

Of late, electron spin resonance (ESR) data have been used for analysis of metal carbonyls or related species. Stolz et al. (25) used ESR to detect $Fe(CO)_4$ fragments in u.v. irradiated solutions of $Fe(CO)_5$; and Rieger and Fraenkel studied the ESR of carbonyl anion radicals (139).

Low-frequency infrared spectral studies have been reported, but the interest lies more in the field of structure determination than analysis. For example, see the report of Edgell et al. (140).

REFERENCES

1. Chatt, J., P. L. Pauson, and L. M. Venanzi, in *Organometallic Chemistry*, H. Zeiss, A.C.S. Monograph No. 147, Reinhold, New York, 1960, Chap. 10.
2. Emeleus, H. J., and J. S. Anderson, *Modern Aspects of Inorganic Chemistry*, 3rd ed., D. Van Nostrand Company, Princeton, New Jersey, 1960.
3. Mattern, J. A., and S. J. Gill, in *Chemistry of the Coordination Compounds*, John C. Bailor, Jr., ed., Reinhold, New York, 1956, Chap. 16.
4. Durrant, P. J., and B. Durrant, *Introduction to Advanced Inorganic Chemistry*, Wiley, New York. 1962.
5. Hileman, J. C., "Carbonyls," in *Kirk-Othmer Encyclopedia of Chemical Technology*, 2nd ed.,, Vol. IV, A.Standen, ed., Interscience, New York, to be published.
5a. E. W. Abel, *Quart Rev.* **17**, 133 (1963).
6. Mond, L., C. Langer, and F. Quincke, *J. Chem. Soc.* **57**, 749 (1890).
7. Mond, L., and F. Quincke, *J. Chem. Soc.* **59**, 604 (1891).
8. Dewar, J., and H. O. Jones, *Proc. Roy. Soc. (London)* **A76**, 558 (1905).
9. Dewar, J., and H. O. Jones, *Proc. Roy. Soc. (London)* **A79**, 66 (1906).
10. Mond, L., H. Hirtz, and M. D. Cowap, *J. Chem. Soc.* **97**, 798 (1910).
11. Job, A., and A. Cassal, *Compt. Rend.* **183**, 58, 392 (1926).
12. Job, A., and J. Rouvillois, *Compt. Rend.* **187**, 564 (1928).
13. Manchot, W., and W. J. Manchot, *Z. Anorg. Allgem. Chem.* **226**, 385 (1936).
14. Hieber, W., and H. Lagally, *Z. Anorg. Allgem. Chem.* **245**, 321 (1940).
15. Hieber, W., and H. Fuchs, *Z. Anorg. Allgem. Chem.* **248**, 256 (1941).
16. Hieber, W., and H. Lagally, *Z. Anorg. Allgem. Chem.* **251**, 96–113 (1943).
17. Hieber, W., and H. Stallman, *Z. Elektrochem.* **49**, 228 (1943).
18. Brimm, E. O., M. A. Lynch, Jr., and W. J. Sesny, *J. Am. Chem. Soc.* **76**, 3831 (1954).
19. Natta, G., R. Ercoli, F. Calderazzo, A. Alberola, P. Corradini, and G. Allegra, *Atti Accad. Naz. Lincei. Rend. Classe Sci. Fis. Mat. Nat.* **27**, 107 (1959); through *Chem. Abstr.*, **54**, 16252 (1960).
20a. Hileman, J. C., D. K. Huggins, and H. D. Kaesz, *J. Am. Chem.* **83**, 2953 (1961); b. K. K. Joshi and P. L. Pauson, *J. Chem. Soc.* 336 (1962).
21a. Kaesz, H. D., and D. K. Huggins, *Can. J. Chem.* **41**, 1250 (1963); b. H. D., Kaesz, private communication, 1964.
22. Nesmeyanov, A. N., K. N. Anisimov, N. E. Kolobova, and I. S. Kolomnikov, *Izv. Akad. Nauk SSSR, Otd. Khim. Nauk* **1963**, 193.
23. Corey, E. R., and L. F. Dahl, *J. Am. Chem. Soc.* **83**, 2203 (1961).
24. Corey, E. R., L. F. Dahl, and W. Beck, *J. Am. Chem. Soc.* **85**, 1202 (1963).
25. Stolz, I. W., G. R. Dobson, and R. K. Sheline, *J. Am. Chem. Soc.* **84**, 3589 (1962).
26. Podall, H. E., *Isolated Bond Strength vs. Molecular Stability for the Metal Carbonyls*, Melpar Inc., Falls Church, Virginia, 1962, pp. 1–10. Presented in part before the Division of Chemical Education, 141st National American Chemical Society Meeting, Washington, D. C., March 21–29, 1962, Abstracts of Papers, P. 8-F.

27. Ormont, B., *Acta Physicochim. URSS* **21**, 413 (1946).
28. Pospekhov, D. A., *Zh. Obshch. Khim.* **18**, 2045 (1948); through *Chem. Abstr.*, **43**, 2535 (1949).
29a. Metlin, S., I. Wender, and H. W. Sternberg, *Nature* **183**, 457 (1959); b. W. Hieber, *Angew. Chem., Intern. Ed. Engl., Sample Issue*, 65–73 (May, 1961).
30a. Grinberg, A. A., B. V. Ptitsyn, F. M. Filinov, and V. N. Lavrent'ev, *Tr. Radievogo Inst. Akad. Nauk SSSR Khim. Geochim.* **7**, 14 (1956); through *Chem. Abstr.* **51**, 17276 (1957); b. R. S. Nyholm, *Proc. Chem. Soc.* **1961**, 273; c. W. Hieber, H. Schulten, and R. Marin, *Z. Anorg. Allgem. Chem.* **240**, 261 (1939).
31. Jolly, W. L., *Synthetic Inorganic Chemistry*, Prentice-Hall, Englewood Cliffs, New Jersey, 1960.
32. *Matheson Gas Data Book*, The Matheson Co., East Rutherford, New Jersey, 1961.
33. Bartocha, B., W. A. G. Graham, and F. G. A. Stone, *J. Inorg. Nucl. Chem.* **6**, 119 (1958).
34. King, R. B., and F. G. A. Stone, *Inorg. Syn.* **7**, 99 (1963).
35. Shore, S. G., *J. Chem. Educ.* **39**, 465 (1962).
36. Ocone, L. R., and J. Simkin, *J. Chem. Educ.* **39**, 463 (1962).
37. Feiser, L. F., *Experiments in Organic Chemistry*, 3rd ed., D. C. Heath, Boston, Massachusetts, 1955.
38. Hileman, J. C., D. K. Huggins, and H. D. Kaesz, *Inorg. Chem.* **1**, 933 (1962).
39. Grosse, A. V., *J. Am. Chem. Soc.* **60**, 212 (1938).
40a. Kobayashi, G., D. K. Huggins, and H. D. Kaesz, private communication, 1962. b. I. Wender, H. W. Sternberg, S. Metlin, and M. Orchin, *Inorg. Syn.* **5**, 190 (1957).
41. Seel, F., "Carbonyls and Nitrosyls," in *Handbook of Preparative Inorganic Chemistry*, Vol. II, G. Brauer, ed., Ferdinand Enke Verlag, Stuttgart, 1962.
42. Arthur, P., Jr., D. C. England, B. C. Pratt, and G. M. Whitman, *J. Am. Chem. Soc.* **76**, 5364 (1954).
43. Reppe, W., and co-workers, *Ann. Chem.* **582**, 116 (1953).
44. Iwanaga, R., *Bull. Chem. Soc. Japan* **35**, 865 (1962).
45. Kirch, L., and M. Orchin, *J. Am. Chem. Soc.* **81**, 3597 (1959).
46. Hieber, W., F. Muhlbauer, and E. A. Ehmann, *Chem. Ber.* **65**, 1090 (1932).
47. Keeley, D. F., and R. E. Johnson, *J. Inorg. Nucl. Chem.* **11**, 33 (1959).
48. Ercoli, R., and F. Barbieri-Hermitte, *Atti Accad. Naz. Lincei Rend. Classe Sci. Fis., Mat. Nat.* **16**, 249 (1954); through *Chem. Abstr.*, **48**, 10408 (1954).
49. Job, A., and A. Cassal, *Bull. Soc. Chim. France* **41**, 814 (1927).
51. Cumming, W. M., J. A. Horn, and P. D. Ritchie, *J. Appl. Chem.* **2**, 624 (1952).
52. Owen, B. B., J. English, Jr., H. G. Cassidy, and C. V. Dundon, *J. Am. Chem., Soc.* **69**, 1723 (1947).
53. Owen, B. B., J. English, Jr., H. G. Cassidy, and C. V. Dundon, *Inorg. Syn.* **3**, 156 (1950).
54. Anisimov, K. N., and A. N. Nesmeyanov, *Compt. Rend. Acad. Sci. URSS* **26**, 58 (1940).
55. Hieber, W., and E. Romberg, *Z. Anorg. Allgem. Chem.* **221**, 321 (1935).
56. Windsor, M. M., and A. A. Blanchard, *J. Am. Chem. Soc.* **56**, 823 (1934).
57. Podall, H. E., J. H. Dunn, and H. Shapiro, *J. Am. Chem. Soc.* **82**, 1325 (1960).
58a. Zakharkin, L. I., V. V. Gavrilenko, and O. Yu. Okhlobystin, Russ. Patent, 111,382 (June 25, 1958); through *Chem. Abstr.*, **52**, 20942 (1958); b. L. I. Zakharkin, V. V. Gavrilenko, and O. Yu. Okhlobystin, *Izv. Akad. Nauk SSSR Otd. Khim. Nauk* **1958**, 100; through *Chem. Abstr.*, **52**, 11736 (1958).

59. Closson, R. D., L. R. Buzbee, and G. G. Ecke, *J. Am. Chem. Soc.* **80,** 6167 (1958).
60. Brimm, E. O., M. A. Lynch, Jr., and W. J. Sesny, U. S. Patent 2,803,525 (August 20, 1957), assigned to Union Carbide Corp.
61a. Natta, G., R. Ercoli, F. Calderazzo, and A. Rabizzoni, *J. Am. Chem. Soc.* **79,** 3611 (1957); b. G. Natta, R. Ercoli, and F. Calderazzo, Ital. Patent 578,713 (July 1, 1958), assigned to "Montecatini" Societa Generale per l'industria mineraria e chimica; through *Chem. Abstr.* **53,** 22784 (1959).
62. Fischer, E. O., and W. Hafner, *Z. Naturforsch.* **10b,** 665 (1955).
63. Fischer, E. O., and W. Hafner, Ger. Patent, 1007,305 (May 2, 1957), assigned to Badische Anilin- und Soda-Fabrik. A.G.
64. Behrens, H., and W. Haag, *Chem. Ber.* **94,** 320 (1961).
65. Podall, H. E., H. B. Prestridge, and H. Shapiro, *J. Am. Chem. Soc.* **83,** 2057 (1961).
66. Nesmeyanov, A. N., E. P. Mikheev, K. N. Anisimov, and N. P. Filimonova, *Russ. J. Inorg. Chem. (English Transl.)* **4,** 1958 (1959).
67. Podall, H. E., U. S. Patent 2,952,521 (Sept., 1960), assigned to Ethyl Corp.
68. Podall, H. E., U. S. Patent 2,952,523 (Sept., 1960), assigned to Ethyl Corp.
69a. Podall, H. E., H. B. Prestridge, and H. Shapiro, *J. Am. Chem. Soc.* **83,** 2057 (1961); b. H. Shapiro and H. E. Podall, *J. Inorg. Nucl. Chem.* **24,** 925 (1963).
70. Nesmeyanov, A. N., K. N. Anisimov, V. L. Volkov, A. E. Friedenberg, E. P. Mikheev, and A. B. Medvedeva, *Russ. J. Inorg. Chem. (English Transl.)* **4,** 1827 (1959).
71a. Hieber, W., and H. Lagally, *Z. Anorg. Allgem. Chem.* **245,** 295 (1940); b. E. Speyer and H. Wolf, *Chem. Ber.* **60,** 1424 (1927); c. R. B. King and F. G. A. Stone, *Inorg. Synth.* **7,** 193 (1962).
72. Hieber, W., *Z. Anorg. Allgem. Chem.* **204,** 165 (1932).
73. Hieber, W., and G. Brendel, *Z. Anorg. Allgem. Chem.* **289,** 324 (1957).
74. Hieber, W., J. Sedlmeier, and R. Werner, *Chem. Ber.* **90,** 278 (1957).
75. Heintzelor, M., Ger. Patent 928,044 (1955), assigned to Badische Anilin- und Soda-Fabrik A.G.
76. Lynch, M. A., Jr., U. S. Patent 2,825,631 (March 4, 1958), assigned to Union Carbide Corp.
77. Closson, R. D., L. R. Buzbee, and G. C. Ecke, *J. Am. Chem. Soc.* **80,** 6167–70 (1958).
78. Hnizda, V., U. S. Patent 2,822,247 (Feb. 4, 1958), assigned to Ethyl Corp.
79. Podall, H. E., U. S. Patent 2,952,522 (Sept., 1960), assigned to Ethyl Corp.
80a. Podall, H. E., and A. P. Giraitis, *J. Org. Chem.* **26,** 2587 (1961); B. J. F. Cordes and D. Neubauer, *Z. Naturforsch.* **17b,** 791 (1962).
81. Addison, C. C., M. Kilner, and A. Wojcicki, *J. Chem. Soc.* **1961,** 4839.
83. Volkov, V. L., E. P. Mikheev, K. N. Anisimov, P. E. Eliseeva, and Z. P. Valueva, *Russ. J. Inorg. Chem. (English Transl.)* **3,** 2433 (1958).
84. D. T. Hurd, U. S. Patent 2,554,194 (May 22, 1951), assigned to General Electric.
85. A. N. Nesmeyanov, E. P. Mikheev, K. N. Anisimov, V. L. Volkov, and Z. P. Valueva, *Russ. J. Inorg. Chem. (English Transl.)* **4,** 503 (1959).
86. Baumgartner, F., and P. Reichold, *Z. Naturforsch.* **16a,** 946 (1961).
87. Behrens, H., and W. Haag, *Chem. Ber.* **94,** 319–9 (1961).
88. Stammerich, H., K. Kawai, and O. Sala, *J. Chem. Soc.* **35,** 2168 (1961).
89. Gilmont, P., and A. A. Blanchard, *Inorg. Syn.* **2,** 242 (1946).
90. Behrens, H., and F. Lohofer, *Chem. Ber.* **94,** 1391 (1961).

91. Behrens, H., H. Zizlsperger, and R. Rauch, *Chem. Ber.* **94**, 1497 (1961).
92. Fischer, E. O., and G. Burger, *Z. Naturforsch.* **17b**, 484 (1962).
93. Behrens, H., and G. von Taueffenbach, *Z. Anorg. Allgem. Chem.* **315**, 259 (1962).
94. Manchot, W., and H. Gall, *Chem. Ber.* **62**, 678 (1929).
95. Blanchard, A. A., J. R. Rafter, and W. B. Adams, Jr., *J. Am. Chem. Soc.* **56**, 16 (1934).
96. Hieber, W., and E. O. Fischer, *Z. Anorg. Allgem. Chem.* **269**, 292 (1952).
97. Hieber, W., and E. O. Fischer, *Z. Anorg. Allgem. Chem.* **269**, 308 (1952).
98. Hieber, W., and R. Bruck, *Z. Anorg. Allgem. Chem.* **269**, 28 (1952).
99a. Hieber, W., *Angew. Chem.* **69**, 469 (1962); b. E. O. Fischer and E. Bockly, *Z. Anorg. Allgem. Chem.* **269**, 308 (1952).
100. Corey, E. R., and L. F. Dahl, *Inorg. Chem.* **1**, 521 (1962).
101. Beck, W., W. Hieber, and G. Braun, *Z. Anorg. Allgem. Chem.* **308**, 23 (1961).
102. Ercoli, R., F. Calderazzo and A. Alberola, *J. Am. Chem. Soc.* **82**, 2966 (1960).
103. Calderazzo, F., and R. Ercoli, *Chim. Ind. (Milan)* **44**, 990 (1962).
104. Werner, R. P. M., and H. Podall, *Chem. Ind. (London)* **1961**, 144.
105. Pruett, R. L., and J. B. Wyman, *Chem. Ind. (London)* **1960**, 119.
106. Nesmeyanov, A. N., K. N. Anisimov, E. P. Mikheev, V. L. Volkov, and Z. P. Valueva, *Russ. J. Inorg. Chem.* **4**, 249 (1959).
107. Corradini, P., *J. Chem. Phys.* **31**, 1676 (1959).
108. Huggins, D. K., and H. D. Kaesz, "Use of Infrared and Raman Spectroscopy in the Study of Organometallic Compounds," in *Progress in Solid State Chemistry*, R. C. Vickery, ed., Pergamon Press, London, 1963.
109. Cable, J. W., R. S. Nyholm, and R. K. Sheline, *J. Am. Chem. Soc.* **76**, 3373 (1954).
110. Bor, G., and L. Marko, *Spectrochim. Acta* **36**, 1105 (1960).
111. Noack, K., *Helv. Chim. Acta* **45**, 1847 (1962).
112. Barraclough, C. C., J. Lewis, and R. S. Nyholm, *J. Chem. Soc.* **1961**, 2582.
113. Jones, L. H., *Spectrochim. Acta* **18**, 329 (1963).
114. Bor, G., *Spectrochim. Acta* **18**, 585–94 (1962).
115. Nyholm, R. S., *Proc. Chem. Soc.* **1961**, 273.
116. Sheline, R. K., and K. S. Pitzer, *J. Am. Chem. Soc.* **72**, 1107 (1950).
117. Edgell, W. F., J. Huff, J. Thomas, H. Lehman, C. Angell, and G. Asato, *J. Am. Chem. Soc.* **82**, 1254 (1960).
118. Hieber, W., W. Beck, and G. Braun, *Angew. Chem.* **72**, 795 (1960).
119. Sheline, R. K., *J. Am. Chem. Soc.* **73**, 1615 (1951).
120. Cotton, F. A., and G. Wilkinson, *J. Am. Chem. Soc.* **79**, 752 (1957).
121. Cotton, F. A., A. D. Liehr, and G. Wilkinson, *J. Inorg. Nucl. Chem.* **2**, 141 (1956).
122. Jones, L. H., *J. Chem. Phys.* **36**, 2375 (1962).
123. Friedel, R. A., I. Wender, S. L. Shufler, and H. W. Sternberg, *J. Am. Chem. Soc.* **77**, 3951–8 (1955).
124. Jones, L. H., *J. Chem. Phys.* **28**, 1215 (1958).
125. Hieber, W., and C. Herget, *Angew. Chem.* **73**, 579 (1961).
126. Hieber, W., J. Peterhans, and E. Winter (in collaboration with W. Beck), *Chem. Ber.* **94**, 2572 (1961).
127. Calderazzo, F., R. Cini, P. Corradini, R. Ercoli, and G. Natta, *Chem. Ind. (London)* **1960**, 500.
129. Rudorff, W., and U. Hoffmann, *Z. Physik Chem. (Leipzig)* **B28**, 351 (1935).

130. Powell, H. M., and R. V. G. Ewens, *J. Chem. Soc.* **1939**, 286.
131. Hanson, A. W., *Acta Cryst.* **15**, 930 (1962).
132. Dahl, L. F., and R. E. Rundle, *J. Chem. Phys.* **26**, 1751 (1957).
133a. Dahl, L. F., E. Ishishi, and R. E. Rundle, *J. Chem. Phys.* **26**, 1750 (1957);
    b. L. F. Dahl and R. E. Rundle, *Acta Cryst.* **16**, 419 (1963).
134. Ladell, J., B. Post, and I. Frankuchen, *Acta Cryst.* **5**, 795 (1952).
135. Corey, E. R., and L. F. Dahl, *Inorg. Chem.* **1**, 521 (1962).
136. Wallach, D., *Acta Cryst.* **15**, 1058 (1962).
137. Evans, D. F., *J. Chem. Soc.* **1959**, 2003.
138. Lundquist, R. T., and M. Cais, *J. Org. Chem.* **27**, 1167 (1962).
139. Reiger, P. H., and G. K. Fraenkel, *J. Chem. Physics* **37**, 2811 (1963).
140. Edgell, W. F., C. C. Helm, and R. E. Anacreon, *J. Chem. Phys.* **38**, 2039 (1963).

CHAPTER 5

Halide and Oxyhalide Complexes of Elements of the Titanium, Vanadium, and Chromium Subgroups

G. W. A. FOWLES

The University, Southampton, England

CONTENTS

The first part of this article reviews the complex halide and oxyhalide compounds formed by elements of the titanium, vanadium, and chromium subgroups, and briefly discusses their properties and structures. In the second part of the article, the available preparative procedures are discussed and compared, and experimental details are given for the preparation of a representative selection of compounds.

I. HALIDE AND OXYHALIDE COMPLEXES OF THE TITANIUM SUBGROUP

Most of the reliable work that has been carried out in this field has been concerned with the hexahalogeno compounds of titanium in the quadri- and ter-valent states and zirconium in the quadrivalent state.

Unlike the elements of the other two subgroups, titanium and zirconium readily form compounds of this type and they have now been well characterized.

A. For the Quadrivalent State

In the case of quadrivalent titanium, complexes of the general formula $M^1_2[TiX_6]$ have been prepared in an anhydrous state for X = F, Cl, and Br, and M^1 = a variety of alkali metals and unipositive organic cations (e.g., pyridinium). The compounds get progressively more difficult to prepare along the series F, Cl, Br, I, and hexaiodo compounds have not been isolated so far. Even though the salts have not been prepared, the anion $[TiI_6]^{2-}$ is believed to be responsible for the dark-red color of the solution formed by titanium(IV) iodide in concentrated hydrogen iodide.

A large number of the hexafluoro salts are known (1) and they are quite stable to heat and are hydrolyzed only very slowly. The hexachloro and hexabromo salts are less well known, and only in recent years have they been prepared in the anhydrous state (2,3). An early report on the bromide (4), for instance, referred to the formation of the diammonium salt as a dihydrate, $(NH_4)_2[TiBr_6] \cdot 2H_2O$, although an examination of the published analytical data shows a Ti:Br ratio of 1:4.5, indicating that the product was hydrolyzed rather than hydrated. Several tetrachlorodibromotitanates have also been prepared (6). Still less is known of the zirconium compounds, especially the bromo derivatives. Thus, a range of the chloro compounds have now been prepared (7), but information on the bromo analogs is limited to the pyridinium and quinolinium compounds (8), and, once again, consideration of the analytical data (Zr:Br = 1.0:4.9 for the pyridinium salt) shows that the products are appreciably hydrolyzed. Table I lists the chloro and bromo

TABLE I
Hexachloro and Bromo Salts of Titanium(IV) and Zirconium(IV)

Compound	Color	M^1
$M^1_2[TiCl_6]$	Yellow	K, Rb, Cs, NH_4, $(CH_3)_4N$, $(CH_3)_3NH$, $(CH_3)_2NH_2$, C_5H_6N, C_9H_8N
$M^1_2[TiCl_4Br_2]$	Red	$(C_2H_5)_4N$, $(C_2H_5)_3NH$, $(C_2H_5)_2NH_2$, C_5H_6N, C_9H_8N
$M^1_2[TiBr_6]$	Dark red	NH_4, C_5H_6N, C_9H_8N
$M^1_2[ZrCl_6]$	White	Na, Rb, Cs, NH_4, $(CH_3)_3NH$, $(C_2H_5)_3NH$, $(CH_3)_2NH_2$, $(C_2H_5)_2NH_2$, $C_2H_5NH_3$, C_5H_6N, C_9H_8N
$M^1_2[ZrBr_6]$		C_5H_6N, C_9H_8N

compounds that have been prepared for quadrivalent titanium and zirconium.

All these hexahalogeno compounds are readily hydrolyzed, and are also ammonolyzed by liquid ammonia (7,9,10); one zirconium–chlorine and two titanium–chlorine or –bromine bonds are ammonolyzed at −33.5°C. The reactions of the bis(alkylammonium) hexachlorozirconate(IV) salts with aliphatic amines have been examined (7), and it has been shown that, while no reaction takes place with tertiary amines, solvolysis occurs slowly with secondary amines, and one chlorine atom is completely replaced in the reactions with ethylamine. The hexachlorozirconate salts are fairly stable to heat and show no sign of decomposition at temperatures of up to 200°C. *in vacuo*.

In the hexahalogeno anions, the metals have an octahedral environment, except for some $[ZrF_6]^{2-}$ salts in the solid state (48). Solutions formed by the chloro compounds in concentrated hydrochloric acid show absorption peaks at 44,800 and 46,500 cm.$^{-1}$ for titanium (11) and zirconium (7), respectively, and these peaks have been attributed to the transfer of the chlorine π electrons to an orbital in the metal atom.

B. For the Ter- and Di-Valent States

Very little is known of the complex halides formed by lower valent zirconium, but over the past ten years much work has been done on the titanium system. Bright and Wurm (12), for instance, have obtained the complex fluorides M_3TiF_6 (with M = Na and K) as violet solids by the electrolytic reduction of the corresponding quadrivalent fluorides. In the case of the chlorides, the hexachloro compounds, M_3TiCl_6, do not appear to have been isolated in a pure state, although early workers reported (13,14) the preparation of the salts $M_2TiCl_5 \cdot H_2O$ (for M = Rb and Cs), in which the molecule of water may complete the octahedral arrangement about the titanium atom. Investigation (15) of the systems $NaCl–TiCl_3$, $KCl–TiCl_3$, and $KBr–TiBr_3$ showed the presence of the hexahalogenotitanate(III) salts in each case. Gruen and McBeth (16) have examined the absorption spectra of Ti^{3+} in a lithium chloride–potassium chloride eutectic at 400°C. and observed a broad band with a maximum at 13,000 cm.$^{-1}$ and a shoulder at 10,000 cm.$^{-1}$; a new band appeared at 7000 cm.$^{-1}$ as the temperature was raised. These authors considered that the spectral results were best explained on the basis of the equilibrium:

$$[TiCl_6]^{3-} \rightleftharpoons [TiCl_4]^- + 2Cl^-$$

The band at lower wavenumbers was attributed to transitions involving the tetrahedral anion.

II. HALIDE AND OXYHALIDE COMPLEXES OF THE VANADIUM SUBGROUP

A. For the Quinquevalent State

Table II summarizes the principal types of halide and oxyhalide complexes that are known. Where compounds of a particular anion have been isolated and characterized, the anion is marked**; where they have been identified only through phase studies or conductivity measurements in suitable solvents, they are marked*. References are given to more recent work, but readers are referred to standard textbooks (e.g., N. V. Sidgwick, *Chemical Elements and Their Compounds*, Oxford, 1950) for details of early work.

TABLE II
Complexes of the Vanadium Subgroup Elements

		Vanadium	Niobium	Tantalum
Fluorides:	$[MF_6]^-$	** (17,18)	**	**
	$[MF_7]^{2-}$	—	**	**
	$[MF_8]^{3-}$		—	**
Oxyfluorides:	$[MOF_4]^-$	**	—	—
	$[MOF_5]^{2-}$	**	**	—
	$[MOF_6]^{3-}$	—	**	**
	$[MOF_7]^{4-}$	—	**	—
Chlorides:	$[MCl_6]^-$	—	*(19–21)	* (21)
Oxychlorides:	$[MOCl_4]^-$	**	**	—
	$[MOCl_5]^{2-}$	—	**	**
Oxybromides:	$[MOBr_4]^-$	—	**	—
	$[MOBr_5]^{2-}$	—	**	—

All three elements form complex fluorides $M^1[MF_6]$, but the analogous chloro compounds have been established only by phase study methods. Thus the $M^1[NbCl_6]$ and $M^1[TaCl_6]$ salts have been shown (21) to be present in the systems formed by niobium(v) chloride and tantalum(v) chloride with chlorides of all the alkali metals (the only exception is in the $LiCl–NbCl_5$ system). The hexafluorovanadates(v) are very easily hydrolyzed and fume in moist air, but the complex fluorides of niobium and tantalum are less easily hydrolyzed. The tantalum compounds are more resistant to hydrolysis than the niobium analogs, and while K_2TaF_7, for instance, can be recrystallized from dilute hydrofluoric acid, K_2NbOF_5 is formed under these conditions.

The structures of some of the complex ions are known, and those of some of the others can be assumed to be closely related. Thus the metal atoms are octahedral in the hexahalogeno anions, and it is very likely that in the $[MOX_5]^{2-}$ anions the oxygen takes the place of one halogen. This is certainly the case for the molybdenum and tungsten compounds that are discussed in the next section. The structure of the

$$
\begin{array}{ccc}
& X & \\
X\diagdown\;|\;\diagup X & & \\
M & & \\
X\diagup\;|\;\diagdown X & & \\
& X &
\end{array}
\qquad
\begin{array}{ccc}
& O & \\
X\diagdown\;||\;\diagup X & & \\
M & & \\
X\diagup\;|\;\diagdown X & & \\
& X &
\end{array}
$$

$[MOX_4]^-$ anions is unknown. It could be a square pyramid (being derived from $[MOX_5]^{2-}$ by the loss of the X^- *trans* to the oxygen without rearrangement) or a trigonal-bipyramid (like the pentahalides in the vapor phase).

The higher fluorides are interesting since they are the simplest compounds in which metals exhibit a coordination number greater than six. The $[MF_7]^{2-}$ anions were originally reported (22) to have a face-centered trigonal-prism structure, but this work needs verifying since the analogous $[ZrF_7]^{3-}$ ion has been shown to have a pentagonal bipyramidal arrangement of fluorine atoms around the metal. The octafluoride complex ion $[TaF_8]^{3-}$ has a square antiprism structure (23).

B. For the Lower Valence States

Table III lists the types of compound that have been reported; the symbols have the same significance as for Table II.

As the information in Table III indicates, little work has been done on the lower oxidation states of niobium and tantalum; the only compounds to be reported so far are the hexachloro derivatives of the quadrivalent elements and $[Nb_2Cl_9]^{3-}$ salts.

TABLE III
Complexes of the Vanadium Subgroup Elements in Oxidation States <5

Oxidation state	Complex anion	Vanadium	Niobium	Tantalum
4+	$[MOF_4]^{2-}$	**	—	—
	$[MCl_6]^{2-}$	* (20)	** (24)	** (24)
	$[MOCl_4]^{2-}$	**	—	—
3+	$[MF_6]^{3-}$	**	—	—
	$[MCl_6]^{3-}$	* (25)	—	—
	$[MCl_4]^-$	** (25,26)		
	$[M_2Cl_9]^{3-}$	** (24)	** (24)	—
	$[MBr_4]^-$	** (26)		
2+	$[MCl_6]^{4-}$	* (25)	—	—
	$[MCl_4]^{2-}$	* (25,27)	—	—
.	$[MCl_3]^-$	* (27)	—	—

With vanadium, on the other hand, quite a number of compounds have been prepared and the presence of other complex anions have been deduced from physical measurements. Thus conductivity studies show (20) that in iodine monochloride solution, potassium chloride and vanadium(IV) chloride react to give the complex $K_2[VCl_6]$. Several chloro anions have been reported for the tervalent element, and spectroscopic investigations (25) on the nature of V^{3+} in a LiCl–KCl eutectic indicate that there is an equilibrium

$$[VCl_6]^{3-} = [VCl_4]^- + 2Cl^-$$

The analogous equilibrium

$$[VCl_6]^{4-} = [VCl_4]^{2-} + 2Cl^-$$

is considered to be present in similar solutions of V^{2+}. Phase studies (27) show the existence of the compounds $KVCl_3$, K_2VCl_4, and $CsVCl_3$ in systems formed by vanadium(II) chloride with potassium and cesium chlorides.

The hexahalogeno and tetrahalogeno derivatives have octahedral and tetrahedral configurations, respectively, but the structures of the oxyhalide complexes and of the $[M_2Cl_9]^{3-}$ and $[VCl_3]^-$ anions are unknown

III. HALIDE AND OXYHALIDE DERIVATIVES OF THE CHROMIUM SUBGROUP

A. For Oxidation States 6+ and 5+

Table IV lists the main types of compounds that have been identified for the elements in these oxidation states.

There have been considerable developments in the characterization of the complex fluorides. The highest fluorides, $M_2^1[MF_8]$ and $M^1[MF_7]$, which are formed by both molybdenum and tungsten, are white solids that are readily hydrolyzed but stable in dry air. Many oxyfluorides have also been prepared. Apart from the chromium derivatives, $M^1[MO_3X]$, no chloro or bromo complexes have been reported for these elements in the 6+ oxidation state.

In the 5+ state, molybdenum and tungsten each form hexafluoro and octafluoro compounds. The former compounds have been examined in some detail. The potassium salts have been shown to have anions with the tetragonal distortion also observed for the niobium anion, $[NbF_6]^-$, namely, an arrangement with four long and two short metal–fluorine bonds. At room temperature, $K[MoF_6]^-$ and $K[WF_6]^-$ have magnetic moments of 1.24 and 0.51 Bohr magnetons (B.M.), respectively; they are both extremely sensitive to traces of moisture, but stable to heat up to 250°C.

The analogous chloro compounds $M^1[MCl_6]$ have not yet been isolated, but a solution of molybdenum(v) chloride in fused potassium chloride has the spectrum predicted for the $[MoCl_6]^-$ anion.

A great deal of work, both preparative and structural, has been done recently on the oxyhalide complexes of quinquevalent molybdenum and tungsten, and interest has been centered especially on the complex anions $[MOX_5]^{2-}$ with $X = Cl$ and Br. Although these anions are essentially octahedral, the multiple metal–oxygen bonding imposes a

TABLE IV

Halide and Oxyhalide Complexes for the Chromium Subgroup Elements in Oxidation States 6+ and 5+

Oxidation state	Anion	Cr	Mo	W
6+	$[MF_8]^{2-}$	—	** (28,29)	** (29,30)
	$[MF_7]^-$	—	** (28)	** (30)
	$[MO_3F]^-$	**	—	—
	$[MO_3F_2]^{2-}$	—	**	**
	$[MO_3F_3]^{3-}$	—	** (31,32)	** (33)
	$[MO_2F_4]^{2-}$	—	**	**
	$[MOF_5]^-$	—	** (34)	** (28)
	$[MO_3Cl]^-$	**	—	—
	$[MO_3Br]^-$	**	—	—
5+	$[MF_6]^-$	—	** (34)	** (34)
	$[MF_8]^{3-}$	—	** (28)	** (30)
	$[MOF_5]^{2-}$	—	**	—
	$[MCl_6]^-$	—	* (35)	—
	$[MOCl_4]^-$	**	** (36)	** (36)
	$[MOCl_5]^{2-}$	**	** (36)	** (36)
	$[MOBr_4]^-$	—	** (36)	** (36)
	$[MOBr_5]^{2-}$	—	** (36)	** (36)

considerable asymmetry with consequent further splitting of the d levels. The spectra of these d^1 systems are accordingly of considerable theoretical interest, and readers are referred to ref. (36) which summarizes the available data and discusses current views.

The chloro and bromo compounds of both metals are green and brown (or yellow-brown), respectively. All are readily hydrolyzed, but stable in air; the analogous chromium compounds are more sensitive to moisture and they also tend to disproportionate and to give chromium(III) derivatives. Salts of the $[MoOCl_5]^{2-}$ anion react with dimethylamine (37) with the solvolysis of one molybdenum–chlorine bond and the formation of $MoOCl_2(NMe_2),NHMe_2$ mixed with dimethyl-amine hydrochloride.

Since the main interest in the $[MOX_4]^-$ compounds is the manner of their preparation, and the reasons why they may be formed in preference to $[MOX_5]^{2-}$ salts, discussion is deferred until the second section of this review. Table V lists the main salts of both types.

TABLE V

Known Salts of Complex Oxyanions of Mo(V) and W(V)

Type of complex	M^1
$M^1{}_2[MoOCl_5]$	NH_4, K, Rb, Cs, C_5H_6N, C_9H_8N, $(CH_3)_3NH$, $(CH_3)_2NH_2$, CH_3NH_3
$M^1[MoOCl_4]$	Rb, Cs, C_5H_6N, C_9H_8N, $(C_2H_5)_2NH_2$
$M^1{}_2[MoOBr_5]$	NH_4, K, Rb, Cs, C_5H_6N, C_9H_8N
$M^1[MoOBr_4]$	C_5H_6N, C_9H_8N
$M^1{}_2[WOCl_5]$	NH_4, Rb, Cs, C_9H_8N, $C_6H_5NH_3$, $(CH_3)_3NH$
$M^1[WOCl_4]$	C_5H_6N, C_9H_8N
$M^1{}_2[WOBr_5]$	NH_4, Rb, Cs
$M^1[WOBr_4]$	C_5H_6N, C_9H_8N, iso-C_9H_8N

B. For Oxidation States 4+, 3+, and 2+

Table VI summarizes the available information in the usual manner. Until recently, the only complexes of the type $M^1{}_2[MX_6]$ that had been prepared for the chromium subgroup elements were the hexa-fluorochromates(IV), but the corresponding fluoro compound of molybdenum has now been made, as have the chloro and bromo derivatives of

TABLE VI

Complexes for the Chromium Subgroup Elements in Oxidation States 4+, 3+, 2+

Oxidation state	Anion	Chromium	Molybdenum	Tungsten
4+	$[MF_6]^{2-}$	**(38)	**(39)	—
	$[MCl_6]^{2-}$	—	**(36,40,41)	**(42,43)
	$[MBr_6]^{2-}$	—	**(43)	**(43)
3+	$[MF_4]^-$	—	**	—
	$[MF_6]^{3-}$	**	—	—
	$[MCl_6]^{3-}$	**	**	—
	$[MBr_6]^{3-}$	—	**	—
	$[M_2Cl_9]^{3-}$	—	**(24)	**
	$[M_3Cl_{14}]^{5-}$	—	—	**(44)
2+	$[MF_3]^-$	**(45)	—	—
	$[MCl_4]^{2-}$	*(16)	—	—
	$[MCl_5]^{3-}$	*(46)	—	—

both molybdenum and tungsten. Table VII lists the compounds that
have now been prepared, together with their colors and magnetic moments (at room temperature).

The magnetic moments are interesting, since they show that as
we go down a group factors such as spin-orbit coupling become increasingly important and the "spin-only" formula cannot be applied.
This point is discussed in detail by Peacock and co-workers (41). The
molybdenum and tungsten compounds are not as easily hydrolyzed as
are other complexes of these elements in the quadrivalent states, and the

TABLE VII
$M^I_2[MX_6]$ Complexes, for M = Mo and W

Complex	M^I	μ(B.M.)	Color
$M^I_2[CrF_6]$	K, Rb	—	Flesh
$M^I_2[MoF_6]$	Na	—	Dark brown
$M^I_2[MoCl_6]$	K	2.28	Dark green
	Rb	2.24	" "
	Cs	2.24	" "
	C_5H_6N	2.36	Yellow
	C_9H_8N	2.30	"
	$(C_2H_5)_2NH_2$	2.30	"
$M^I_2[MoBr_6]$	Rb	2.18	Olive green
	Cs	2.08	" "
$M^I_2[WCl_6]$	K	1.43	Red
	Rb	1.48	"
	Cs	1.48	"
	$(CH_3)_3NH$	1.55	Yellow
	$(C_2H_5)_3NH$	1.34	Brown
	$(C_2H_5)_2NH_2$	1.59	Pink
$M^I_2[WBr_6]$	K	1.50	Olive green
	Rb	1.43	" "
	Cs	1.72	Grass green

rubidium and cesium salts are less easily hydrolyzed than the others.
They are insoluble in solvents with which they do not react, except for
such halogen solvents as iodine monochloride. The dipotassium hexa-
chloromolybdate and hexabromomolybdate complexes react with liquid
ammonia in an unexpected manner, there being apparently disproportionation and ammonolysis, with the formation of the insoluble complexes of tervalent molybdenum, $K_3Mo_2X_9$, and soluble ammonobasic
molybdenum(v) halides. Because of the insolubility of the hexachloro
complexes, spectra measurements have been made only on the solids,
and so far it is not certain whether the three bands observed in the
21,000–28,000 cm.$^{-1}$ region are ligand-field or charge-transfer bands.

The formation of complex anions of the type $[MX_6]^{3-}$ by tervalent chromium and molybdenum is well established, but tungsten only appears to form complex anions containing two or more tungsten atoms bridged by chlorine atoms. In the $[W_2Cl_9]^{3-}$ anion, for instance, three chlorines bridge two WCl_3 groups. $M_3[W_2Cl_9]$ has been found (47) to react with pyridine and aniline to give substituted derivatives (e.g., $py_3W_2Cl_6$) that are nonelectrolytes.

Gruen and McBeth (16) have examined the spectra of both Cr^{3+} and Cr^{2+} in a LiCl–KCl eutectic and found only negligible changes over a temperature range 400–1000°C.; these measurements show the presence of only one species for each system, namely, $[CrCl_6]^{3-}$ for Cr^{3+} and $[CrCl_4]^{2-}$ for Cr^{2+}, and there seem to be no octahedral–tetrahedral equilibria as observed for the solutions formed by titanium and vanadium halides. From a phase study of the system formed by NaCl and $CrCl_2$, Shiloff reports (46) the formation of Na_3CrCl_5.

The crystal structure of the complex fluoride formed by divalent chromium, $KCrF_3$, is interesting, since it contains each chromium surrounded by six fluorines in a distorted octahedron. The observed four long and two short chromium–fluorine bond distances are predicted by simple ligand-field theory.

IV. METHODS USED FOR THE PREPARATION OF HALIDE AND OXYHALIDE COMPLEX SALTS

There are two general methods that can be used for the preparation of these complex compounds:

(1) Direct reaction, usually at high temperatures, in the absence of a solvent, e.g.,

$$3M^1X + MX_3 \rightarrow M^1_3[MX_6]$$

(2) Direct reaction in a suitable solvent.

(1) In one sense this is rather an artificial division, because reactions coming under the first heading can be regarded as making use of one of the reactants (usually M^1X) as a solvent, i.e., fused salt. It is useful to consider the first type separately, however, because such preparations involve particular experimental problems. Thus there is a general difficulty in ensuring the formation of pure compounds, since one or other of the reactants is usually in slight excess, and cannot be removed in many cases (i.e., if it is either involatile or insoluble in solvents with which the product does not react). In some instances the reactants can, of course, be mixed in exactly the correct ratio without undue experimental difficulty, but it is essential to ensure that the product is a

true compound rather than a mixture. In the general equation given above, for example, we have supposed that three moles of the alkali metal halide react with one mole of the transition metal halide to give one mole of the complex $M^1{}_3[MX_6]$, but the product might be a mixture (e.g., $2M^1X + M^1[MX_4]$), so that some further identification (e.g., phase study, x-ray) is a necessary safeguard before this method can be used for the preparation of compounds that have not been characterized previously. Its use is generally restricted to the preparation of alkali metal salts, because at the high temperatures involved, ammonium and substituted ammonium halides dissociate giving ammonia or amines which may attack some of the transition metal halides. This may be illustrated by reference to $(NH_4)_2[TiCl_6]$ which cannot be prepared by the direct reaction of NH_4Cl and $TiCl_4$ in a sealed tube; some ammonolysis invariably takes place.

The fused salt procedure has been used to prepare alkali metal salts of several complex halide anions, e.g. $[TiCl_6]^{2-}$ and $[ZrCl_6]^{2-}$, but in many cases the compounds can be prepared more easily and in a purer state by methods making use of a solvent. It is a particularly valuable approach for the preparation of complex halides of some elements (e.g., Nb and Ta) in their lower oxidation states, however, because the simple halides are often insoluble in the more usual solvents at room temperature but may dissolve in fused alkali metal halides. Furthermore, the simple halides are not always prepared very easily in a pure state, but they can be prepared *in situ* in a fused salt medium.

(2) Methods using solvents are widely used, and the solvent may be the aqueous hydrohalic acid, ethanolic hydrogen chloride or bromide, or one of a variety of aprotic solvents. The choice depends upon the particular complex that is being prepared. Table VIII outlines the various methods that are in common use, although it must be stressed that this list is not exhaustive by any means, and, as we shall see shortly, there are alternative methods for many of the compounds.

One general point stands out immediately. If we consider the complexes formed by the elements in their higher oxidation states, then, whereas titanium and zirconium form halide complexes in aqueous acid, elements of the vanadium and chromium subgroups give only oxyhalide complexes under the same conditions (niobium and tantalum do, however, form fluoro complexes in aqueous hydrofluoric acid).

We will now consider the various groups, in turn, in a fairly general manner, the exact preparative details being given in the final section.

We shall not, however, discuss the preparation of the fluoro complexes to any significant extent, because either they are very straightforward and prepared in aqueous hydrofluoric acid by very well established pro-

TABLE VIII
Summary of Preparative Methods

Element	Oxidation state	Complex	Reaction medium	Reactants, etc.
Ti, Zr	4+	$M^1_2[MF_6]$	Aqueous HF	—
Ti: Zr	4+	$M^1_2[MX_6]$ ($X = Cl, Br$)	Aqueous HX^a or ethanolic HX^b	$MX_4 + M^1X$
Ti	3+	$K_3[TiF_6]$	NaCl–KCl	K_2TiF_6—electrolytic reduction
V	5+	$M^1[VF_6]$	BrF_3	$VF_5 + M^1F$
Nb; Ta	5+	Fluorides and oxyfluorides	Aqueous HF	—
V; Nb; Ta	5+	Oxychlorides and oxybromides	Aqueous HX	—
Nb; Ta	4+	$M^1_2[MCl_6]$	None	$MCl_5 + M^1 +$ M^1Cl
V	3+	$M^1[VX_4]$ ($X = Cl, Br$)	CH_3CN	$VX_3 + M^1X$
Mo; W	6+	Fluorides	IF_5 or SO_2	$MF_6 + M^1F$
Cr; Mo; W	6+	Oxyfluorides	Aqueous HF	—
Mo; W	6+	Fluorides	IF_5 or SO_2	$MF_6 + M^1I$
Cr; Mo; W	5+	$M^1_2[MOX_5]$ ($X = Cl$ and Br)	Aqueous HX	$MX_5 + M^1X$
Mo; W	5+	$M^1[MOX_4]$ ($X = Cl$ and Br)	(i) Aqueous HX (ii) SO_2	$MX_5 + M^1X$
Cr	4+	$M^1_2[MF_6]$	None	$CrCl_3 + M^1Cl +$ F_2
Mo	4+	$Na_2[MoF_6]$	SO_2	$MoF_6 + KI$
Mo	4+	$M^1_2[MoCl_6]$	(i) $CHCl_3$	$MoCl_4,2RCN +$ M^1Cl
			(ii) ICl	$MoCl_5 + M^1Cl$
	4+	$M^1_2[MoBr_6]$	IBr	$MoBr_3 + M^1Br$
W	4+	$M^1_2[WCl_6]$	ICl	$WCl_6 + M^1Cl$
	4+	$M^1_2[WBr_6]$	None	$WBr_6 + M^1I$
Mo	3+	$M^1_3[MoX_6]$ ($X = Cl, Br$)	Aqueous HX	Reduction of molybdate
W	3+	$M^1_3[W_2Cl_9]$	Aqueous HX	Reduction of tungstate

a For alkali metal salts.
b For substituted ammonium salts, etc.

cedures, or else they require very specialized apparatus suitable for use with highly reactive interhalogen compounds such as BrF_3. In the latter case, interested readers are referred to recent reviews (48,49), and to experimental procedures already published in *Inorganic Syntheses* (50).

A. Titanium and Zirconium

The chlorides and bromides are prepared either in aqueous or ethanolic hydrogen halide. An aqueous medium is used for preparing the ammonium, rubidium, cesium, pyridinium, and quinolinium salts, and ethanol for the alkyl-substituted ammonium salts. For a preparation in aqueous acid, the transition metal halides and the alkali metal halide are dissolved in separate amounts of aqueous hydrogen halide (of maximum molarity), the solutions mixed, cooled to $0°C$, and hydrogen chloride or bromide gas passed through Ammonium chloride is not very soluble in concentrated hydrochloric acid, and in this instance the initial solution is made with $8M$ acid; because of the insolubility of sodium chloride in hydrochloric acid, it is difficult to prepare sodium salts by this procedure The other salts precipitate quite readily, and they are perfectly stable when wet with concentrated acid, but great care is necessary in the final stages of isolating the dry anhydrous salt. Thus if the last traces of acid are removed directly *in vacuo*, hydrogen halide is first removed, leaving the constant boiling acid, and in such concentrations of hydrogen halide the complex anions are hydrolyzed.

This difficulty is overcome if the aqueous acid is washed out first. This is done in the case of the chloro complexes by means of a solution of 5% thionyl chloride in ether, followed by anhydrous ether. Thionyl chloride is particularly suitable for this purpose since its hydrolysis products are SO_2 and HCl, neither of which can interact with the hexachloro anions; it is, however, unwise to use neat thionyl chloride as the wash liquid, since the last traces of it are not easily removed from the complex salt. Presumably thionyl bromide could be used in the same manner for the final isolation stages of the complex bromides, but because this is not readily available bromine has been used instead.

In the case of the titanium complex $(NH_4)_2TiBr_6$, bromine has been used as the solvent as well as wash liquid, since titanium(IV) bromide can be prepared *in situ* by the direct reaction of titanium metal sponge with bromine.

Preparations in ethanol are carried out in a manner exactly analogous to those in water, except that thionyl chloride is no longer suitable as a final wash liquid, since some displacement of the amine cation seems to take place. Successive washings with anhydrous ethanol are satisfactory.

B. Vanadium, Niobium, and Tantalum

Very little work has been carried out recently on the oxychloride and oxybromide complexes of these elements. With vanadium, for instance,

the oxychloride complexes for the quinquevalent and quadrivalent
element, namely, $M^1[VOCl_4]$ and $M^1_2[VOCl_4]$, are said to be very soluble, so much so in fact that only the pyridinium and quinolinium salts
have been reported; these were prepared in ethanol.

The complex oxychlorides and oxybromides of quinquevalent niobium
and tantalum have also received little attention in the past fifty years.
The niobium series $M^1_2[NbOCl_5]$, with M^1 = Rb and Cs, is prepared
quite readily by mixing aqueous hydrochloric acid solutions of the
alkali metal chloride and niobium(v) chloride or niobium(v) oxytrichloride. Pyridinium and quinolinium salts of the $[NbOCl_4]^-$ anion
have been prepared by the analogous reactions both in aqueous and ethanolic hydrogen chloride. It is interesting to note here that when two
such series of salts are formed, the alkali metals usually form the M^1_2-
$[MOX_5]$ type and the larger organic cations form the second $M^1[MOCl_4]$
type.

Very little is known about the complex halides of niobium and tantalum in their lower valence states, although it is clear that such compounds
can be prepared by the fusion technique. Thus the interaction of $NbCl_4$
and KCl or NaCl produces the complexes $M^1_2[NbCl_6]$, and the same
compounds can be prepared by reduction of a solution of $NbCl_5$ in M^1Cl
by the alkali metal.

One or two salts of the tetrahedral tetrachloro- and tetrabromo-
vanadate(III) anions have been prepared by reaction in methyl cyanide.
Both the trichloride and tribromide dissolve in methyl cyanide (and
other alkyl cyanides) under reflux, although the complexes that are
formed, $VX_3,3RCN$, are only slightly soluble in excess of the alkyl
cyanide at room temperature. In the presence of alkylammonium halides such as tetraethylammonium chloride, the salt is formed:

$$VCl_3 \xrightarrow[Et_4NCl]{MeCN} [Et_4N][VCl_4], 2MeCN \longrightarrow [Et_4N][VCl_4]$$

Unfortunately, the exact details of this preparation have not yet been
published, and cannot be given in the next section.

C. Chromium, Molybdenum, and Tungsten

For the elements in the quinquevalent state, there are two series of
oxychloride and oxybromide complexes, as with niobium: $M^1_2[MOX_5]$
and $M^1[MOX_4]$. Table V gives the list of salts that have already
been prepared. All the complexes of the first type are prepared in aqueous hydrochloric or hydrobromic acid, by reacting a solution of M^1X
with one containing either molybdenum or tungsten in the quinquevalent state. The latter solutions can be prepared by electrolytic reduc-

tion of molybdate or tungstate solution, but this requires the construction of a suitable cell, and so the methods given here involve the use of molybdenum(v) chloride (which is commercially available) or potassium oxalato-dioxytungstate(v) (which is easily prepared). The chromium compounds are somewhat less stable, but can be prepared in a reasonably pure form by somewhat analogous methods; CrO_3 is the chromium source, and it is reduced to the quinquevalent state by the glacial acetic acid–hydrogen chloride solvent.

In the case of tungsten, this aqueous acid procedure gives complexes of the second type in some instances where large cations are used, e.g., (C_9H_8N) $[WOCl_4]$, (C_5H_6N) $[WOBr_4]$, (C_9H_8N) $[WOBr_4]$, (iso-C_9H_8-N) $[WOBr_4]$.

In general, however, the second type of complex is best prepared using liquid sulfur dioxide as a solvent. Thus $MoCl_5$ reacts with $M^I Cl$ in liquid sulfur dioxide to give the $M^I[MoOCl_4]$ salts; $MoCl_5$ is first solvolyzed to give the oxytrichloride $MoOCl_3$, and this reacts to give the complex salt. The pyridinium and quinolinium salts can also be prepared by treating the $M^I_2[MoOCl_5]$ salts with liquid sulfur dioxide, when a breakdown takes place: $M^I_2[MoOCl_5] \rightarrow M^I Cl + M^I[MoOCl_4]$. There is an interesting side reaction in the preparation of the rubidium and cesium salts, since two products are formed in either case, one the expected oxychloride derivative $M^I[MoOCl_4]$ and the other the hexachloromolybdate(IV) salt $M^I_2[MoCl_6]$. It is not clear just how this reduction does occur, and why it is only observed in reactions involving the rubidium and cesium chlorides; nevertheless, yields in excess of 30% are obtained.

The alkali metal derivatives $M^I_2[MoCl_6]$ have also been prepared by reaction of $MoCl_5$ with $M^I Cl$ in iodine monochloride. The easiest way of making the corresponding salts of organic cations is to react $M^I Cl$ with the complex compound $MoCl_4 \cdot 2C_3H_7CN$ in chloroform solution; the n-propyl cyanide complex is very readily prepared by treating $MoCl_5$ with excess of the cyanide, when the molybdenum is reduced quantitatively to the quadrivalent state.

The alkali metal salts of the analogous hexachlorotungstate(IV) anion can again be made by reactions in iodine monochloride; this and the analogous bromo compounds are also made by the reaction between WCl_6 or WBr_6 and $M^I I$ in the absence of a solvent.

The complexes of molybdenum and tungsten in the tervalent state, $M^I_3[MoX_6]$ and $M^I_3[W_2Cl_9]$, can be prepared fairly easily by the reduction of concentrated hydrochloric acid solutions of the sexivalent elements (i.e., MoO_3 and $M^I_2WO_4$). The molybdenum reduction is usually carried out electrolytically and the tungstate is reduced by tin. Since experimental details are available in *Inorganic Syntheses* (51), they will not be repeated in the next section.

V. PREPARATIVE DETAILS

A. Titanium Subgroup

1. Cs_2TiCl_6 (2)

$TiCl_4$ and $CsCl$ are dissolved in separate amounts of concentrated hydrochloric acid. The concentrations are not critical, but the $CsCl$ solution is usually a saturated one. Portions of the two separate solutions are mixed to give $TiCl_4:CsCl$ slightly greater than $1:2$. The solution is then cooled to $0°C$. and saturated with HCl gas. The pale-yellow crystals formed are filtered and washed successively with (*a*) concentrated HCl, (*b*) a 5% solution of thionyl chloride in diethyl ether, and (*c*) dry diethyl ether. The washing with the thionyl chloride solution is continued until there is no further effervescence. Finally, the crystals are pumped for several hours under a vacuum of around 10^{-3} mm. During the final stages (i.e., ether washing and pumping), it is important that the crystals are not exposed to moist air; therefore, it is an advantage if the filtration vessel forms a part of the vacuum system so that the washing and pumping can be carried out without interruption to transfer the sample.

The ammonium, potassium. rubidium, pyridinium, and quinolinium salts are made in a directly analogous manner. The only variation in procedure is in the preparation of the solution of M^1Cl in hydrochloric acid; where it is difficult to obtain a sufficiently concentrated solution in $11M$ hydrochloric acid, the acid is diluted.

2. $[(CH_3)_3NH]_2TiCl_6$ (2)

The procedure is similar to that given under (*1*) for Cs_2TiCl_6 except that anhydrous ethanol saturated with HCl is used as the solvent. $(CH_3)_3NHCl$ is made *in situ* by dissolving the amine in ethanol and passing in HCl gas. The product is washed with anhydrous ethanol.

Other alkylammonium salts, e.g., $[(CH_3)_2NH_2]_2TiCl_6$ and $[CH_3NH_3]_2$-$TiCl_6$ are made in the same way.

3. $[(CH_3)_3NH]_2ZrCl_6$ (7)

This and a wide range of alkylammonium salts are made as for the titanium analogs. The white crystalline hexachlorozirconates do not always precipitate immediately, however, and it is then necessary to allow the solution to stand over night in a refrigerator.

4. $[C_5H_5NH]_2TiBr_6$ (2)

TiBr$_4$ (0.01 mole) and C$_5$H$_5$N (0.02 mole) are dissolved separately in 48% aqueous HBr, the solutions mixed, cooled to 0°C., and saturated with HBr gas. The dark-red crystals are filtered, washed successively with anhydrous bromine (distilled from phosphoric oxide prior to use) and anhydrous diethyl ether, and finally pumped *in vacuo*

Since TiBr$_4$ is not readily available, some workers may prefer to use freshly precipitated titanium "hydroxide" dissolved in HBr; furthermore, it may be worthwhile to make thionyl bromide for use as a wash liquid (in ether solution) rather than to use bromine.

B. Vanadium Subgroup

1. $Cs_2[NbOCl_5]$

This, and the analogous rubidium compound, may be prepared by the method described for the preparation of Cs$_2$TiCl$_6$ (A1), using either NbOCl$_3$ or NbCl$_5$ as the starting material. The corresponding bromides can be made in aqueous HBr (see method A4).

C. Chromium Subgroup

1. $Cs_2[CrOCl_5]$

CrO$_3$ is dissolved in cold glacial acetic acid saturated with hydrogen chloride and the solution is allowed to stand for about 30 min. Addition of a solution of CsCl in 11M HCl results in the immediate precipitation of the complex salt as a fine red powder. This product is filtered and washed as was Cs$_2$TiCl$_6$ (method A1).

2. $Cs_2[MoOCl_5]$ and $Cs_2[WOCl_5]$ (36)

These, and the other salts listed in Table V are prepared by method A1, but using MoCl$_5$ and WCl$_5$ as sources of molybdenum and tungsten, respectively. The commercial MoCl$_5$ is somewhat contaminated by hydrolysis products, but this is not important since in any case the hydrochloric acid hydrolyzes the chloride to give the [MoOCl$_5$]$^{2-}$ anion. Since WCl$_5$ is not readily available, some workers may prefer to use K[WO$_2$(C$_2$O$_4$)].xH$_2$O as the source of quinquevalent tungsten. This compound is prepared as follows:

An aqueous solution containing 6.5 g. potassium oxalate, 3 g. oxalic acid, and 1.3 g. potassium tungstate is heated to boiling, and then reduced by granulated tin at 80°C. As this reduction takes place, the solution changes color through blue, to green, and finally red. When reduction is complete (usually about 1 hr.), H_2S is passed through the solution to precipitate the tin as tin sulfide; the solution is filtered into 500 ml. ethanol. Red crystals slowly deposit, and these are filtered off and washed with ethanol.

3. $(C_5H_6N)[MoOCl_4]$ (36)

C_5H_6NCl (0.022 mole) and $MoCl_5$ (0.01 mole) are placed in an ampoule attached to a vacuum line, an excess of sulfur dioxide is condensed in, and the ampoule sealed off. The complex salt precipitates and can be filtered off on the vacuum line (the ampoule being cooled before it is opened), then washed with $CHCl_3$ to remove the excess C_5H_6NCl. The quinolinium salt is made in an analogous manner, but the diethylammonium salt is very soluble in liquid SO_2 so that in this case the solvent is evaporated and the residue treated with $CHCl_3$ to remove the excess hydrochloride.

The rubidium and cesium salts are also soluble in SO_2, but they are contaminated with the $M^1{}_2MoCl_6$ salts which are insoluble. In these cases, $MoCl_5$ is used in slight excess, the reaction mixture is filtered, the SO_2 is evaporated from the filtrate, and the residue remaining is treated with $CHCl_3$ to remove unchanged $MoCl_5$.

4. $(C_5H_6N)_2[MoCl_6]$ (36)

$MoCl_4, 2C_3H_7CN$ is first prepared by sealing $MoCl_5$ with an excess of C_3H_7CN in an ampoule for several days. The ampoule is then opened, attached to a vacuum line, and the excess of C_3H_7CN removed. Extraction of the residue with dry benzene yields $MoCl_4, 2C_3H_7CN$. $CHCl_3$ solutions of $MoCl_4, 2C_3H_7CN$ (0.01 mole), and C_5H_6NCl (0.02 mole) are mixed, and the bright-yellow precipitate is filtered off under vacuo, washed with $CHCl_3$, and then pumped for several hours. The quinolinium and diethylammonium salts can be prepared in the same way.

The rubidium and cesium salts are best prepared as described in method C3.

5. $Cs_2[WCl_6]$ (43)

CsI (0.01 mole) is finely ground and placed in a tube in an oven for several hours at 130°C.; WCl_6 (0.0055 mole) is then added (either in a dry box or by conventional vacuum-line procedures) and the tube sealed. The tube is kept at 130°C. for 3 days, and then opened and attached to a vacuum line. I_2 and the excess of WCl_6 are removed by heating the sample to around 300°C. The potassium and rubidium salts are prepared in a similar manner.

Iodine monochloride can be used as a supporting solvent if desired.

6. $Cs_2[WBr_6]$ (43)

This and the analogous potassium and rubidium salts are prepared by method C5, but using CsI and WBr_6 as reactants.

REFERENCES

1. Sidgwick, N. V., *The Chemical Elements and Their Compounds*, Oxford Univ. Press, New York, 1950, p. 645.
2. Fowles, G. W. A., and D. Nicholls, *J. Inorg. Nucl. Chem.* **18**, 130 (1961).
3. Wernet, J., *Z. Anorg. Allgem. Chem.* **272**, 279 (1953).
4. Jander, J., *Z. Anorg. Allgem. Chem.* **294**, 181 (1958).
5. Rosenheim, A., and O. Schütte, *Z. Anorg. Allgem. Chem.* **150**, 69 (1925).
6. Bye, J., and W. Haegi, *Compt. Rend.* **236**, 381 (1953).
7. Drake, J. E., and G. W. A. Fowles, *J. Inorg. Nucl. Chem.* **18**, 136 (1961).
8. Rosenheim, A., and P. Frank, *Chem. Ber.* **38**, 812 (1905).
9. Fowles, G. W. A., and D. Nicholls, *J. Chem. Soc.* **1961**, 95.
10. Drake, J. E., and G. W. A. Fowles, *J. Less-Common Metals* **3**, 149 (1961).
11. Jorgensen, C. K., *Absorption Spectra and Chemical Bonding in Complexes*, Pergamon, New York, 1962, p. 284.
12. Bright, N. F. H., and J. G. Wurm, *Can. J. Chem.* **36**, 615 (1958).
13. Stähler, A., *Chem. Ber.* **37**, 4405 (1904).
14. Stähler, A., *Chem. Ber.* **38**, 2619 (1905).
15. Ehrlich, P., G. Kupa, and K. Blankenstein, *Z. Anorg. Allgem. Chem.* **299**, 213 (1959).
16. Gruen, D. M., and R. L. McBeth, *Seventh International Conference on Co-ordination Chemistry*, Butterworths, London, 1963, p. 27; *Nature* **194**, 468 (1962).
17. Emeléus, H. J., and V. Gutmann, *J. Chem. Soc.* **1949**, 2979.
18. Sharpe, A. G., and A. A. Woolf, *J. Chem. Soc.* **1951**, 798.
19. Palkin, A. P., and N. D. Chiranov, *Zh. Neorgan. Khim.* **4**, 898 (1959).
20. Gutmann, V., *Z. Anorg. Allgem. Chem.* **264**, 151 (1951).
21. Huber, K., E. Just, E. Neuenschwander, M. Studer, and B. Roth, *Helv. Chim. Acta*, **41**, 2411 (1958).
22. Hoard, J. L., *J. Am. Chem. Soc.* **61**, 1252 (1939).
23. Hoard, J. L., W. J. Martin, M. E. Smith, and J. F. Whitney, *J. Am. Chem. Soc.* **76**, 3820 (1954).
24. Ijdo, D. J. W., private communication.
25. Gruen, D. M., and R. L. McBeth, *J. Phys. Chem.* **66**, 57 (1962).
26. Nyholm, R. S., *Croat. Chem. Acta* 1961, **33**, 157.
27. Seifert, H. J., and P. Ehrlich, *Z. Anorg. Allgem. Chem.* **302**, 284 (1959).
28. Hargreaves, G. B., and R. D. Peacock, *J. Chem. Soc.* **1958**, 4390.
29. Cox, B., D. W. A. Sharp, and A. G. Sharpe, *J. Chem. Soc.* **1956**, 1242.
30. Hargreaves, G. B., and R. D. Peacock, *J. Chem. Soc.* **1958**, 2170.
31. Schmitz-Dumont, O., and I. Heckmann, *Z. Anorg. Allgem. Chem.* **267**, 277 (1952).
32. Schmitz-Dumont, O., and P. Opgenhoff, *Z. Anorg. Allgem. Chem.* **275**, 21 (1954).
33. Schmitz-Dumont, O., I. Bruns, and I. Heckmann, *Z. Anorg. Allgem. Chem.* **271**, 347 (1953).
34. Hargreaves, G. B., and R. D. Peacock, *J. Chem. Soc.* **1957**, 4212.
35. Horner, S., and S. Y. Tyree, private communication.

36. Allen, E. A., B. J. Brisdon, D. A. Edwards, G. W. A. Fowles, and R. G. Williams, *J. Chem. Soc.* **1963,** 4649.

37. Edwards, D. A., and G. W. A. Fowles, unpublished observations.

38. Huss, E., and W. Klemm, *Z. Anorg. Allgem. Chem.* **262,** 25 (1950).

39. Edwards, A. J., and R. D. Peacock, *Chem. Ind. (London)* **1960,** 1441.

40. Allen, E. A., D. A. Edwards, and G. W. A. Fowles, *Chem. Ind. (London)* **1962,** 1026.

41. Edwards, A. J., R. D. Peacock, and A. Said, *J. Chem. Soc.* **1962,** 4643.

42. Brisdon, B. J., G. W. A. Fowles, and B. P. Osborne, *J. Chem. Soc.* **1962,** 1330.

43. Kennedy, C. D., and R. D. Peacock, *J. Chem. Soc.* **1963,** 3392.

44. Laudise, R. A., and R. C. Young, *J. Am. Chem. Soc.* **77,** 5288 (1955).

45. Edwards, A. J., and R. D. Peacock, *J. Chem. Soc.* **1959,** 4126.

46. Shiloff, J. C., *J. Phys. Chem.* **64,** 1566 (1960).

47. Jonassen, H. B., S. Cantor, and A. R. Tarse, *J. Am. Chem. Soc.* **78,** 271 (1956).

48. Sharpe, A. G., *Advan. Fluorine Chem.* **1,** 29 (1959).

49. Peacock, R. D., in *Progress in Inorganic Chemistry,* Vol. 2, F. A. Cotton, ed., Interscience, New York, 1960, p. 193.

50. Priest, H. F., in *Inorganic Syntheses,* Vol. 3, L. F. Audrieth, ed., McGraw-Hill, New York, 1950, p. 171; D. R. Martin, in *Inorganic Syntheses,* Vol. 4, J. Bailar, ed., McGraw-Hill, New York, 1953, p. 133.

51. Lohmann, K. H., and R. C. Young, in *Inorganic Syntheses,* Vol. 4, J. Bailar, ed., McGraw-Hill, New York, 1953, p. 97; R. A. Laudise, and R. C. Young, in *Inorganic Syntheses,* Vol. 6, E. G. Rochow, ed., McGraw-Hill, New York, 1960, p. 149.

CHAPTER 6

Anhydrous Metal Nitrates

C. C. ADDISON AND N. LOGAN

University of Nottingham, Nottingham, England

CONTENTS

I. INTRODUCTION

Nitrates of many metals are known in the form of their hydrates, but, until recent years, relatively few anhydrous nitrates had been prepared. The anhydrous nitrate of each of the alkali metals is well known; in Group II of the Periodic Table the anhydrous nitrates of calcium, strontium, and barium are readily available by removal of water from the hydrate. The anhydrous nitrates of silver, lead, and thallium(I) are also familiar compounds. It is a surprising fact, however, that until quite recently the anhydrous nitrates of the majority of the metals had not been isolated. So far as preparative work is concerned, the major obstacle has been the persistent use of aqueous systems; attempts to remove water from the hydrates commonly result in hydrolysis with the formation of hydroxide nitrates, hydroxides, or oxides, and the evolution of nitric acid. This experience may have been responsible for the general belief that anhydrous nitrates, particularly those of the transition elements, would be unstable compounds. In fact, many such compounds

141

have recently been isolated and shown to have high thermal stability. Their preparation is a direct consequence of the more extensive use of nonaqueous solvents in preparative inorganic chemistry.

Preparative work in this field has been stimulated by the observation that the physical properties and chemical reactivity of some of the anhydrous nitrates are quite different from what might be expected on the basis of the known chemistry of the metal nitrate hydrates. The chemistry of these compounds has been discussed in detail elsewhere (1), but the salient features will be mentioned here since they have direct repercussions on the preparative techniques employed. It has now been realized that the bond between a metal atom and a nitrate group can be strongly covalent, and that the nitrate group can act as a unidentate, bidentate, or bridging ligand. In the simple nitrates $M(NO_3)_2$, where $M = Mn$, Ni, Cu, or Zn, the nitrate groups are covalently bonded, but the crystal structure of copper(II) nitrate (2) indicates that in each of the nitrate groups more than one oxygen atom is involved in covalent bonding. It may well be proven that the unidentate nitrate group is only to be found in complexes such as $Mn(CO)_5NO_3$ (3) and Me_3SnNO_3 (4). In the former, the metal-to-nitrate bond can survive many substitution reactions in which the carbonyl groups are displaced (5). Bidentate bonding occurs in the ion $[UO_2(NO_3)_3]^-$ (6) and in the di- and tri-hydrates of uranyl nitrate (7); in the anion $[M(NO_3)_4]^{2-}$, where $M = Cu$ or Ni, the metal atom appears to be 6-coordinate in the complex, so that bidentate bonding must again occur (8). The bridging of the metal atoms by nitrate groups is a prominent feature in the structure of basic beryllium nitrate (9) and solid copper(II) nitrate (2).

The most important physical property resulting from this strong covalent bonding is the pronounced volatility of certain anhydrous nitrates. For example, the compounds $Cu(NO_3)_2$ (10), $Ti(NO_3)_4$ (11, 16,59), $Zr(NO_3)_4$ (12), and $Be_4O(NO_3)_6$ (9) can exist in the vapor state and, in consequence, can be purified by sublimation. The chemical properties which are introduced by covalent bonding must also be recognized when considering preparative methods. Many covalent nitrates [e.g., $Cu(NO_3)_2$, $Zn(NO_3)_2$] are highly soluble in polar organic solvents; indeed, copper nitrate is more soluble in ethyl acetate than in water, and when dissolved cannot be crystallized again from ethyl acetate (13). In aqueous solution, normal dissociation into metal cations and nitrate ions usually, but not invariably, occurs. On solution of anhydrous beryllium nitrate in water, about 10% of the nitrate groups appear as nitrite ions in the solution (14). Some nitrates in which covalent bonding is particularly strong can react (even explosively) with organic compounds. Thus, copper nitrate will react vigorously with diethyl ether (15) and nitromethane (13,15), and titanium tetranitrate reacts with dodecane to give alkyl nitrate, nitroalkane, and a carboxylic acid (16).

The methods devised for the preparation of anhydrous metal nitrates must take into account the physical and chemical properties of each individual nitrate, and these may vary widely. The methods almost invariably involve nonaqueous solvents. Some solvents have been little used to date, but will be mentioned briefly at this stage since they may have potential value in special cases. Metathetic reactions between silver nitrate and various metal halides dissolved in acetone give solutions of the metal nitrates, but the solid products do not appear to have been isolated (17). Displacement reactions between metals and solutions of silver nitrate in acetone, phenyl cyanide, or liquid ammonia give the metal nitrates in the form of their adducts with the solvent (18). Attempts to prepare anhydrous nitrates using pure nitric acid as reaction medium have generally been unsuccessful; the reaction of pure nitric acid with a number of anhydrous metal chlorides gave hydrated nitrates or oxides as products (19,20). Under the influence of anhydrous hydrogen chloride and the metal ions, the nitric acid appears to dissociate into its anhydride and water, and the latter is removed by precipitation of the hydrated metal nitrate. The use of molten salts as preparative media has not yet been studied to any extent; the main difficulties are the isolation of the product, and the high temperatures involved. When mercury(II) nitrate dihydrate is dissolved in molten mercury(II) bromide (21), the water of crystallization evaporates at the temperatures involved (238–320°C.). Again, the lanthanum oxide La_2O_3 reacts with molten ammonium nitrate at 170°C. Excess ammonium nitrate may be removed by volatilization, leaving anhydrous lanthanum nitrate (22). This method merits further investigation.

The preparative methods which have the widest general application involve the use of dinitrogen tetroxide or dinitrogen pentoxide and their derivatives, and it is these methods which will be discussed in more detail.

II. REACTIONS USING DINITROGEN TETROXIDE

Liquid dinitrogen tetroxide is available commercially in some countries and is readily purified by distillation. It can conveniently be prepared in pure form in the laboratory by thermal decomposition of lead nitrate (23,24). It has a convenient liquid range (m.p. − 11.2°C., b.p. 21.15°C.) and reactions can usually be carried out at room temperature. The most important requirement in using the liquid is that it must be protected from moisture. Containing vessels should be dried at 100°C. before use. The liquid may be handled in a glass apparatus, and reactions can be carried out by pouring solutions from one vessel into another through connecting tubes. Closed vessels so connected are vented to the atmosphere through side-tubes carrying phosphoric oxide supported on glass wool. Some dinitrogen tetroxide escapes to the atmosphere in this way, so the operations should be conducted in a fume cupboard. One of the most sensitive tests for accidental contamination by moisture is to freeze the tetroxide; the pure tetroxide freezes to a colorless glass, whereas traces of moisture give rise to dinitrogen trioxide which imparts a green color to the frozen tetroxide. Before carrying out a reaction, it is usually advisable to cool the liquid about 10 °C. below its boiling point, otherwise the heat of reaction may cause vigorous boiling and the evolution of large quantities of toxic fumes from the apparatus.

The self-ionization of the liquid is represented by the equilibrium

$$N_2O_4 \rightleftharpoons NO^+ + NO_3^-$$

This has been confirmed by several physical methods (25–27) and is entirely consistent with chemical evidence. It follows that in this liquid the nitrate ion has the same significance as has the hydroxyl ion in water. The self-ionization is very small in the pure liquid, but is enhanced by dilution with polar solvents, and the preparations described below are subdivided according to the liquid medium used. Reactions normally employ the metals themselves, the anhydrous metal halides, the metal oxides, or the metal carbonyls.

A. Dinitrogen Tetroxide Alone

Metal halides undergo solvolysis by dinitrogen tetroxide more readily in the presence of donor solvents, and examples will be given at a later stage. Nitrates of some lanthanide elements have been prepared by direct reaction of dinitrogen tetroxide with the metal oxide in a sealed bomb (28), but this technique has not been widely used because of the high temperatures involved. Reactions of the tetroxide alone with metal carbonyls and with metals are important, and an example of each reaction is given below.

1. Reaction with Metal Carbonyls

The pure liquid reacts with all metal carbonyls which have been tested, the rate of reaction depending on the physical state of the metal carbonyl and also on temperature (3). Thus, the liquid carbonyls (nickel tetracarbonyl and iron pentacarbonyl) react violently; the solids dimanganese decacarbonyl and dicobalt octacarbonyl react readily with liquid dinitrogen tetroxide at room temperature, but reaction is negligible at 0 °C.

a.   Anhydrous Manganese(II) *Nitrate.*   The expensive dimanganese decacarbonyl would not normally be used as starting material for the preparation of the anhydrous nitrate, but the following preparation illustrates the type of reaction taking place in these systems.

Dinitrogen tetroxide (6 ml.) is added to dimanganese decacarbonyl (1 g.) contained in a glass tube vented to atmosphere through a phosphoric oxide guard tube, at room temperature (20°C.).   The carbonyl initially dissolves in the tetroxide without reaction, but after about 3 min. a vigorous reaction commences, producing a dark-green cloudy solution:

$$Mn_2(CO)_{10} + 4N_2O_4 \rightarrow 2Mn(NO_3)_2 + 4NO + 10CO$$

The vigorous reaction subsides after 30 min., but evolution of gas and precipitation of a yellow solid continues for 20 hr.   After a further 24 hr., the precipitate is collected on the sintered glass plate of a filter tube which is protected by phosphoric oxide guard tubes.   The precipitate is washed with dinitrogen tetroxide and dried in a stream of nitrogen.   The yellow product is the adduct $Mn(NO_3)_2 \cdot N_2O_4$, from which the anhydrous nitrate may be isolated as a white powder by heating at 100°C. under vacuum (10^{-2} mm.) for 12 hr. (5).

If the dimanganese decacarbonyl is allowed to react at 0°C. with dinitrogen tetroxide which is diluted with light petroleum, the compound $Mn(CO)_5NO_3$ may be isolated from the solution (3).   This is the only metal carbonyl nitrate prepared to date, and all other metal carbonyls tested give the simple nitrates as products.

Manganese(II) nitrate can be prepared in larger quantities by solution of powdered manganese metal in a mixture of dinitrogen tetroxide and ethyl acetate.   The subsequent treatment of the solution is the same as that described below for the isolation of the compound $Be(NO_3)_2$.   This, (and all other nitrates discussed in this article) is strongly deliquescent, and contact with atmosphere must be avoided.

2. Reaction with Metals

The reaction $M + xN_2O_4 \rightarrow M(NO_3)_x + xNO$ has been observed to occur with the alkali metals (29,30), zinc (31), and mercury (32), but most other metals [including barium, strontium, and calcium (14)] do not react in the massive state.   Due to the low dielectric constant of the pure liquid tetroxide (2.42), no simple inorganic salts are soluble, and the reaction product forms as an insoluble coating on the metal surface.   The efficiency of the reaction, therefore, depends on the cohesive state of the metal nitrate produced.   Reaction of sodium is vigorous on first immersion in the liquid, but sodium nitrate is highly cohesive and the reaction soon stops completely. Reaction with zinc goes to completion, and has been studied in some detail.

a. Anhydrous Zinc Nitrate. A block of zinc (5 g.) having a freshly filed surface is immersed in liquid dinitrogen tetroxide (10 ml.) at about 20°C. Within 10 min., the metal is covered by a thin layer of white, finely crystalline solid; nitric oxide is evolved, and the liquid in immediate contact with the metal is colored green by dinitrogen trioxide. As the coating becomes thicker, the crystals of the adduct $Zn(NO_3)_2 \cdot 2N_2O_4$ are more clearly defined, and can easily be removed from the metal surface by mild agitation. About 1–2 g. of product form in 24 but the reaction is more difficult to control. After the removal of excess metal, the dinitrogen tetroxide is allowed to evaporate, when the product is obtained as a free-flowing, finely crystalline powder. The dinitrogen tetroxide is weakly bonded, and is evolved on heating the powder. The removal of about 10% of the combined tetroxide causes the solid to change to liquid; the phase change results, no doubt, from depression of the melting point of the original adduct by the zinc nitrate produced. The reaction

$$Zn(NO_3)_2 \cdot 2N_2O_4 \leftrightarrows Zn(NO_3)_2 + 2N_2O_4$$

is reversible; complete removal of the tetroxide is effected by heating under reduced pressure for 6 hr. at 100°C. Anhydrous zinc nitrate remains as a free-flowing white powder (31).

Dinitrogen trioxide produced in solution during the zinc metal–dinitrogen tetroxide reaction has a measurable effect on the reaction rate (33). If the block of metal is undisturbed, the reaction rate gradually increases, but if the zinc block is stirred in the liquid the rate is decreased. This is in contrast to the usual behavior of heterogeneous metal–liquid reactions, where a concentration gradient of active species is set up at the surface and stirring increases the rate of reaction. The observed behavior is attributed to the presence of dinitrogen trioxide at the metal surface. This molecule is more polar than is the tetroxide, and it will, therefore, increase both the dielectric constant of the medium and the concentration of the active NO^+ species; its presence at the surface will enhance reaction rates, and it will be dispersed on stirring. The trioxide takes no direct part in the reaction, since its self-ionization into NO^+ and NO_2^- ions would inevitably lead to the formation of metal nitrite as the product. The rate of solution of zinc is proportional to $[N_2O_3]^{1/2}$; at 0°C. the rate is eight times as great in a solution containing 40% dinitrogen trioxide as in dinitrogen tetroxide alone, but the product contains no trace of nitrite.

B. Dinitrogen Tetroxide with an Inert Diluent

The term "inert solvent" is used in this context to signify a solvent which does not interact with dinitrogen tetroxide to any appreciable extent, and which does not form a molecular addition compound with the tetroxide in the liquid or solid states. The advantages to be gained by using dinitrogen tetroxide–inert solvent mixtures rather than the tetroxide alone will be illustrated by reference to nitromethane, which has been most widely used for this purpose. Nitromethane gives stable mixtures with dinitrogen tetroxide, and is miscible in all proportions. No addition compounds are formed, and the phase diagram shows a simple eutectic liquidus curve (34). Anhydrous metal nitrates formed as reaction products dissolve in the mixture, and can be recovered by crystallization. The self-ionization of the tetroxide is enhanced by the high dielectric constant ($\epsilon = 37$) of nitromethane, so that the mixture is more reactive than the tetroxide alone. Preparations using mixtures with nitromethane have not yet been studied extensively, since the same products are often more rapidly available using mixtures of the tetroxide with donor solvents such as ethyl acetate (Sec. II.C). However, in cases where the metal nitrate forms a strong molecular addition compound with the donor solvent, it is preferable to use nitromethane. For example, uranium metal reacts as follows:

$$U + N_2O_4 + EtOAc \rightarrow UO_2(NO_3)_2 \cdot 2EtOAc$$

$$U + N_2O_4 + MeNO_2 \rightarrow UO_2(NO_3)_2 \cdot N_2O_4 \rightarrow UO_2(NO_3)_2$$

The anhydrous nitrate cannot be isolated from the ethyl acetate adduct (32); in nitromethane solution the 1:1 dinitrogen tetroxide adduct is formed, from which anhydrous uranyl nitrate can be obtained by careful heating.

Figure 1 shows the variation in rate of reaction of copper, zinc, and uranium with the composition of the mixture. The rate curve for copper (which does not react with the tetroxide alone) passes through a sharp maximum at about 85 mole-% of nitromethane (35). The electrical conductivity of the mixtures varies in a very similar way, so that the fastest reaction occurs at compositions at which the number of NO^+ ions is greatest. The basic reaction is, therefore, considered (35) to be

$$Cu + NO^+ \rightarrow Cu^+ + NO$$

and the close correlation between conductivity and reaction rate holds for copper because this metal has a stable Cu(I) state. The dissolution rate reflects the transfer of a single electron; oxidation to Cu(II) then occurs, but it is not a rate-determining process. From the solutions, the

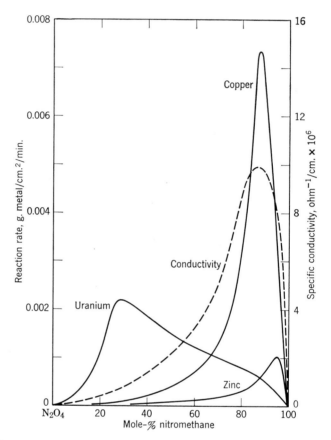

Fig. 1. Rates of reaction of metals with N_2O_4–CH_3NO_2 mixtures at 15°C. (Stirring rate 400 r.p.m. The uranium values shown are magnified by a factor of 50.)

compound $Cu(NO_3)_2 \cdot N_2O_4$ may be crystallized. This is usually prepared from ethyl acetate mixtures (Sec. II.C) and need not be discussed further at this stage. The rate curve (Fig. 1) for zinc resembles that for copper, though the maximum is somewhat displaced. The rates of reaction of uranium in Figure 1 are magnified to show that the maximum in the rate curve now bears no relation to the conductivity values. Neither zinc nor uranium possesses a stable monovalent state; the metal–NO+ reaction is of a higher order, and reaction rate will not, therefore, depend directly on the NO+ ion concentration. The product of reaction with zinc is the same compound, $Zn(NO_3)_2 \cdot 2N_2O_4$, as has already been described. The preparation of anhydrous uranyl nitrate will now be described to illustrate the use of inert diluents. There is an additional feature of interest in that the dinitrogen tetroxide in the adduct UO_2-$(NO_3)_2 \cdot N_2O_4$ is very strongly bonded and can only be removed under strictly controlled conditions.

1. Anhydrous Uranyl Nitrate

A piece of freshly filed uranium metal is immersed in excess of a mixture of dinitrogen tetroxide and nitromethane containing about 30 mole-% of nitromethane. After some hours at room temperature, the liquid turns green in color, due partly to the dinitrogen trioxide formed and partly to $U(IV)$ species in solution. There is no retardation due to insoluble reaction products, and the reaction proceeds steadily. When the metal is dissolved this green color disappears, since $U(IV)$ in solution which is oxidized to $U(VI)$ by the dinitrogen tetroxide can no longer be replaced from the metal. Excess dinitrogen tetroxide is then added, and a light-yellow microcrystalline powder which has the composition UO_2-$(NO_3)_2 \cdot N_2O_4$ is thrown out of solution. Alternatively, the compound can be obtained in the form of large yellow crystals by allowing dinitrogen tetroxide vapor to dissolve slowly in a concentrated solution of the yellow powder in nitromethane (36).

The same compound has been prepared by reaction of uranium oxides with dinitrogen tetroxide in an autoclave (37); uranium trioxide obtained by heating $UO_4 \cdot 2H_2O$ for 150 hr. at 500°C. is recommended for this reaction (38), but when the oxide is sealed together with dinitrogen tetroxide in glass ampoules the reaction still requires 48 to 72 hr. to go to completion. Uranium tetrachloride, uranium pentachloride, and uranyl chloride are quickly converted into $UO_2(NO_3)_2 \cdot N_2O_4$ when added to dinitrogen tetroxide–nitromethane mixtures (39).

The temperature at which the tetroxide addition compound must be heated to obtain anhydrous uranyl nitrate is highly critical. The pure nitrate is obtained by heating the adduct at 163°C. for 20 hr. under a pressure of 0.1 mm. Hg (36), or at 165°C. for 110 min. and a vacuum of 10^{-5} mm. Hg (38). If the temperature is varied by more than a few degrees from 163 to 165°C., the anhydrous nitrate is not obtained. For example, samples heated at 160°C. for long periods retain small amounts of dinitrogen tetroxide, and samples heated at 176°C. contain uranium trioxide impurity (36).

C. Dinitrogen Tetroxide with Donor Solvents

1. Nature of Liquid Mixtures

The dinitrogen tetroxide molecule is electron deficient, and so is the NO^+ ion formed by self-ionization. When the tetroxide is mixed with organic solvents which possess an atom capable of donating electrons, partial electron transfer occurs. When such mixtures are cooled, molecular addition compounds can be crystallized. In the liquid phase, both N_2O_4 and NO^+ species are associated with the donor solvent, and the ultraviolet spectra of these mixtures indicate that discrete molecules of an addition complex are not present (40). The liquid mixtures are best represented by the equilibrium

$$[(Don)_nNO]^+ + NO_3^- \rightleftharpoons n(Don) + N_2O_4 \rightleftharpoons (Don)_n \cdot N_2O_4$$

where (Don) represents the electron donor solvent. In preparative work it is the weaker donor solvents which are important. The species $(Don)_n \cdot NO^+$ then behaves chemically in the same manner as the NO^+ ion alone; the concentration of the latter is thus enhanced both by dielectric constant changes, and by the formation of the coordinated nitrosonium ions. A surprisingly large number of organic solvents will form stable liquid mixtures with dinitrogen tetroxide. They include carboxylic acids (41), esters (41,42), anhydrides (42), ketones (41,42), ethers (43), nitriles (41,42), nitrosamines (41,44), and sulfoxides (45). Table I lists some examples of solvents which may be used in this way.

The chemical reactivity of dinitrogen tetroxide is stimulated to a much greater extent by donor solvents than by inert solvents. This is shown clearly by the reactions of these solutions with metals. Figure 2

TABLE I

Some Solvents Which Show Electron Donor Properties Toward Dinitrogen Tetroxide

Solvent	Solid adduct with N_2O_4		Donor atom
	Mole ratio	M.P., °C.	
Diethylnitrosamine	2:1	−37	Nitrogen
Ethylphenylnitrosamine	2:1	−12	Nitrogen
Methyl cyanide	2:1	−42	Nitrogen
Methyl cyanide	1:1	−39	Nitrogen
Benzyl cyanide	2:1	−43	Nitrogen
Diethylsulfoxide	1:1	+13	Sulfur
Di-isopropylsulfoxide	1:1	−10	Sulfur
Di-isopropylsulfoxide	2:1	− 3	Sulfur
Acetic acid	2:1	+ 2	Oxygen
Ethyl acetate	2:1	−65 and −72	Oxygen
Ethyl benzoate	2:1	−12	Oxygen
Acetic anhydride	1:1	−44	Oxygen
Diethylether	2:1	−74.8	Oxygen
1,4-Dioxan	1:1	+45.2	Oxygen
1,3-Dioxan	1:1	+ 2.0	Oxygen
Ethyleneglycol diethyl ether	1:1	−60	Oxygen

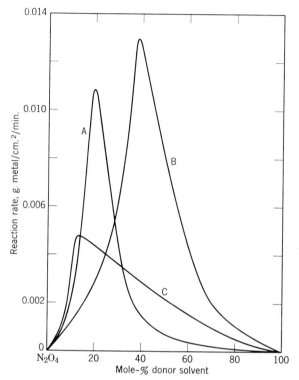

Fig. 2. Rates of reaction with N_2O_4- donor solvent mixtures (15°C. and 400 r.p.m.)
A. Copper-ethyl acetate; *B*. copper-diethyl ether; *C*. zinc-ethyl acetate.

shows three typical solution rate curves, and many other metals for
which detailed reaction rates have not been measured are known to
behave in similar fashion. With the nitromethane solutions, reaction
rates for copper and zinc metals (Fig. 1) were very small over the 0–50
mole-% nitromethane concentration range, but reaction rates increase
rapidly on addition of relatively small quantities of ether or ethyl acetate
(Fig. 2). This reflects the presence of $(Don)_nNO^+$ ions in solution,
which can take part in electron-transfer reactions of the type:

$$M + (Don)_nNO^+ \rightarrow M^+ + n(Don) + NO$$

In all donor solvent systems, some restrictive effect which reduces reaction rate appears at higher solvent concentrations. The onset of this
effect is sudden, and gives rise to the sharp maxima in the rate curves
(Fig. 2). This restrictive effect is concerned with the ability of the molecules present in the liquid to remove the metal ions from the surface
once they are formed, either by solvation or coordination. Further con-

sideration of this aspect is not relevant here.   Cyanides and nitrosamines
give extreme examples of the effect of donor solvents (Fig. 3).

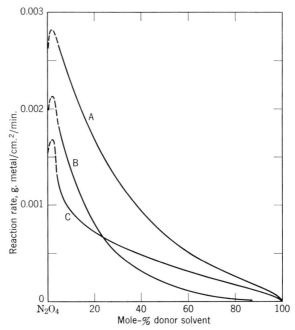

Fig. 3. Rates of reaction with mixtures of liquid N_2O_4 and nitrogen donor solvents
(15°C. and 400 r.p.m.).   *A*. Copper-diethylnitrosamine;   *B*. copper-phenyl cyanide;
C. zinc- benzyl cyanide.

Reaction rates rise immediately to their maximum value on the addition
of only a few mole-% of the solvent, and then fall continuously throughout the concentration range (35).

It will be seen from the above that there is a wide variation in the reactivity of metals with dinitrogen tetroxide–donor solvent mixtures, depending on the metal and the solvent used.   A variety of products may
also be obtained by crystallization from solutions of metals in these
mixtures.

2. Nature of Products

These may be classified as follows:

(*a*) Metal nitrate
(*b*) Metal nitrate–dinitrogen tetroxide adduct
(*c*) Metal nitrate–solvent adduct
(*d*) Metal nitrate–N_2O_4–solvent adduct.

The particular product obtained depends on the metal, the solvent, and on the concentration of the latter. Some typical examples are given in Table II. Whether or not the metal nitrate can be obtained from the adduct with solvent or dinitrogen tetroxide depends on the extent to which the solvent is coordinated to the metal; this obviously depends upon the metal and the solvent, and each adduct is an individual problem.

TABLE II

Influence of Donor Solvent on Initial Product from
Metal–Dinitrogen Tetroxide Reactions

Metal	Reaction medium: N_2O_4 with	Initial reaction product	Ref.
Cadmium	Ethyl acetate	$Cd(NO_3)_2$	46
Manganese	Ethyl acetate	$Mn(NO_3)_2 \cdot N_2O_4$	5
Cobalt	Ethyl acetate	$Co(NO_3)_2 \cdot 2N_2O_4$	47
Uranium	Ethyl acetate	$UO_2(NO_3)_2 \cdot 2EtOAc$	32
Uranium	Diethylnitrosamine	$UO_2(NO_3)_2 \cdot 2Et_2N \cdot NO$	36
Nickel	Methyl cyanide	$Ni(NO_3)_2 \cdot 2MeCN$	48
Indium	< 20 mole-% Methyl cyanide	$In(NO_3)_3 \cdot 2N_2O_4$	49
Indium	> 20 mole-% Methyl cyanide	$In(NO_3)_3 \cdot 2MeCN$	49
Bismuth	Dimethylsulfoxide	$Bi(NO_3)_3 \cdot 3Me_2SO$	50
Copper	5–30% Phenyl cyanide	$Cu(NO_3)_2 \cdot 2PhCN \cdot 4N_2O_4$	51
Copper	Ethyl acetate	$Cu(NO_3)_2 \cdot N_2O_4$	10

Discussion so far has been concerned with reactions with metals. An additional advantage in the use of these mixtures in preparative work arises from the fact that many metal halides which do not react with dinitrogen tetroxide alone will undergo solvolysis in these mixtures. Ethyl acetate and methyl cyanide have been most widely used for this purpose. Three preparations involving halides or metals will now be described which illustrate the variation in possible techniques.

a. Nickel(II) Nitrate. Metallic nickel does not react at room temperature with dinitrogen tetroxide alone, nor its mixtures with ethyl acetate or methyl cyanide. Anhydrous nickel chloride is also unattacked by liquid dinitrogen tetroxide at room temperature. Methyl cyanide is, therefore, added to a suspension of nickel chloride (1 g.) in dinitrogen tetroxide (10 ml.). The quantity of methyl cyanide added is not critical, but reaction is more rapid if relatively small amounts are used (see Fig. 2). An immediate reaction takes place, and nitrosyl chloride is evolved. The nickel chloride then dissolves slowly to give a clear green solution, and the reaction is complete after 48 hr. at room temperature. Excess dinitrogen tetroxide and methyl cyanide are removed by evacuation at room temperature; the color of the solution changes from green to blue, and eventually a deep-blue solid of composition $Ni(NO_3)_2 \cdot 3MeCN$ is obtained (48). Prolonged vacuum treatment converts this to the turquoise-green solid $Ni(NO_3)_2 \cdot 2MeCN$. The latter decomposes under vacuum at 170 °C. to leave the anhydrous $Ni(NO_3)_2$ as a pure, lime-green powder.

Nickel nitrate may also be obtained by solvolysis of nickel chloride in dinitrogen tetroxide–ethyl acetate mixtures, when the adduct $Ni(NO_3)_2 \cdot N_2O_4$ is the initial product (48).

b. *Copper*(II) *Nitrate.* A freshly abraded strip of pure copper (10 g.) is added to a mixture of dinitrogen tetroxide (20 ml.) and dried ethyl acetate (20 ml.) in a glass tube closed by a ground joint and phosphoric oxide guard tube. Nitric oxide is evolved vigorously and the reaction is completed in 4 hr. at room temperature. Excess liquid dinitrogen tetroxide is then added to the blue solution to cause complete precipitation of $Cu(NO_3)_2 : N_2O_4$ as a blue-green finely crystalline solid. This is collected on the sintered glass plate of a filter tube, washed several times with small portions of liquid dinitrogen tetroxide, and transferred (in the dry box) to a 17×3 cm. tube where excess dinitrogen tetroxide is removed under vacuum.

Thermal decomposition of the adduct is effected by immersing the tube in an oil bath, and maintaining the vacuum while raising the temperature slowly to 120 °C. Anhydrous copper(II) nitrate then remains as a blue amorphous powder (10). The compound can be obtained in crystalline form by sublimation. A layer (0.5 cm.) of clean, dry glass wool is packed on to the surface of the powder in the bottom of the tube; the tube is so arranged that the oil level of the bath is about 3 cm. above the surface of the glass wool and is again evacuated. On raising the bath temperature to 200 °C., the copper nitrate rapidly sublimes, the sublimate collecting on the walls of the tube about 0.3 cm. above the oil level. Alternatively, the rich blue crystals may be collected on a cold finger inserted in the tube to this level. Using a vacuum of 10^{-3} mm., the sublimation of 10 g. of copper nitrate is complete in about 4 hr., leaving a residue of only about 5% of the original material. Slower sublimation is possible at much lower temperatures; at 130 °C., 100 mg. quantities have been collected in 2 hr., and smaller quantities at temperatures as low as 100 °C. Copper nitrate condenses, at first, in the form of fine, needle-like crystals which are suitable for x-ray crystallographic study. At a later stage in the sublimation, these thicken to give a solid mass of crystals (10).

c. *Beryllium Nitrate.* There are two complicating factors in the preparation of this compound: firstly, the high solubility of beryllium nitrate in ethyl acetate, and, secondly, the ready decomposition of

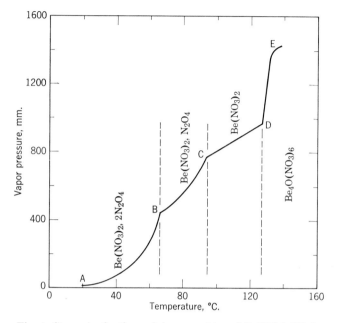

Fig. 4. Stages in the thermal decomposition of $Be(NO_3)_2 \cdot 2N_2O_4$.

beryllium nitrate to the volatile basic nitrate. Anhydrous beryllium chloride (2 g., which may be prepared *in situ* from the metal and chlorine) is dissolved in ethyl acetate (50 ml.) and dinitrogen tetroxide is added dropwise. Each addition gives rise to a vigorous effervescence of nitrosyl chloride, and addition of the tetroxide is continued until this effervescence no longer occurs. Alternatively, beryllium chloride is suspended in liquid dinitrogen tetroxide, and ethyl acetate added dropwise; effervescence ceases when all the beryllium chloride has dissolved. From the clear solution obtained by either method, final traces of nitrosyl chloride are rapidly removed under vacuum. If attempts are made at this stage to precipitate beryllium nitrate by addition of dinitrogen tetroxide, very large quantities of the latter are required. The solution is therefore warmed (40–50°C.) and most of the ethyl acetate removed under vacuum until a viscous gum remains. Liquid dinitrogen tetroxide (100 ml.) is then added to the gum when pale-yellow crystals of the solvate $Be(NO_3)_2 \cdot 2N_2O_4$ gradually separate (14).

In order to determine conditions for dissociation of this adduct to the simple nitrate, the compound has been heated in an evacuated vessel, and the pressure of evolved vapor determined over a temperature range by means of a glass spiral gauge (14). The results (Fig. 4) give a clear indication of the steps involved in thermal decomposition. The successive loss of the two molecules of dinitrogen tetroxide is indicated by the break B in the curve in the temperature range AC. Part CD of the curve follows the gas laws and beryllium nitrate is, therefore, stable up to 125 °C. When heated above 125 °C. (DE, Fig. 4), rapid decomposition occurs; nitrogen dioxide is evolved, and basic beryllium nitrate, $Be_4O(NO_3)_6$, condenses in the form of cubic crystals on the cool parts of the vessel.

III. REACTIONS WITH DINITROGEN PENTOXIDE

Dinitrogen pentoxide has not yet been used to the same extent as has the tetroxide for the preparation of anhydrous metal nitrates. In practice, its manipulation in the pure state is more difficult since at any temperature above 0 °C. it slowly decomposes into the tetroxide and oxygen, and the range of solvents in which it can be dissolved without reaction is not so extensive as for dinitrogen tetroxide. The pure compound is conveniently prepared by oxidizing a stream of nitrogen dioxide with excess oxygen which has passed through an ozonizer, and condensing or dissolving the product (53). Pure dinitrogen pentoxide melts at 41 °C. In the liquid state or in solution, it is believed to undergo self-ionization to the NO_2^+ and NO_3^- ions which are characteristic of the solid state. In many cases it does not appear necessary to employ the pure compound, and a number of successful preparations have been carried out using the impure product obtained by distillation from a mixture of nitric acid and phosphoric oxide; this usually contains a little nitric acid and dinitrogen tetroxide.

A. Dinitrogen Pentoxide Alone

Some typical preparations involving dinitrogen pentoxide will now be described to illustrate the methods used and a further example will be referred to in Sec. IV.

1. Zirconium(IV) Nitrate

This may be prepared by direct reaction between anhydrous zirconium tetrachloride and dinitrogen pentoxide

$$ZrCl_4 + 4N_2O_5 \rightarrow Zr(NO_3)_4 + 4NO_2Cl$$

and purification of the pentoxide does not appear to be necessary (12). The covalent tetranitrate is volatile, and may be separated by sublimation. Excess of dinitrogen pentoxide, prepared from fuming nitric acid and phosphorus pentoxide, is distilled directly on to freshly prepared anhydrous zirconium tetrachloride contained in a vessel cooled in liquid nitrogen. The mixture is allowed to warm slowly to room temperature, and is then heated to 30°C. The impure dinitrogen pentoxide melts at this temperature and is allowed to reflux for 15 min., during which time reaction occurs and nitryl chloride is evolved. The system is evacuated for an hour (0.1 mm. pressure) at room temperature, when excess pentoxide is removed, and the yellow solid remaining has the composition $Zr(NO_3)_4$, 0.4 N_2O_5, 0.6 N_2O_4. The coordinated nitrogen oxides are removed by heating for 4 hr. at 100°C./0.01 mm., after which the anhydrous $Zr(NO_3)_4$ slowly sublimes on to a water-cooled cold finger in the form of colorless crystals.

2. Titanium(IV) Nitrate

Dinitrogen pentoxide appears to be one of the few reagents capable of removing water from a hydrated metal nitrate without decomposition of the anhydrous nitrate. The method has been used to prepare anhydrous nitrates of copper, hafnium, zirconium, and titanium, and may prove to be a good general method for the preparation of anhydrous nitrates (16).

In the preparation of the titanium compound, a solution of titanium-(III) chloride is treated with ammonium hydroxide, and the hydrated titanium hydroxide which is precipitated is dissolved in fuming nitric acid. The solution is gently evaporated to dryness to give hydrated titanium(IV) nitrate. An excess of dinitrogen pentoxide (prepared from fuming nitric acid and phosphorus pentoxide) is distilled directly on to the hydrated nitrate, which is contained in a vessel cooled in liquid nitrogen. Subsequent treatment of this mixture is similar to that described above for the zirconium compound, except for the fact that titanium(IV) nitrate is much more volatile. After evacuation at room temperature, the temperature is raised to about 36°C., when the tetranitrate begins to sublime under a vacuum of 0.03 mm., and condenses as large cubic crystals on a water-cooled cold finger. Sublimation is rapid at higher temperatures, and at 100°C. (0.02 mm.) as much as 2 g. may be sublimed in an hour (16).

3. Uranyl Nitrate

An interesting solid–solid reaction occurs on mixing uranium trioxide and pure dinitrogen pentoxide (53). The active form (37) of UO_3 (0.0028–0.0036 mole), screened through a 325-mesh sieve, is stirred with pure dinitrogen pentoxide (0.09 mole) in powder form at 0–10°C.

Stirring should be vigorous, and a magnetic stirrer is convenient for this purpose. A stream of ozone is passed over the mixture for the duration of the experiment. A light-orange mixture is originally formed but within 10 min. the mixture suddenly turns bright yellow in color without any evident heat of reaction. Within a further 1–2 min., a finely divided, crystalline free-flowing yellow powder is obtained. Excess N_2O_5 is removed at 25–30°C. and at 10^{-4} mm. pressure when the product has the composition $UO_2(NO_3)_2 \cdot N_2O_5$. It seems probable that reaction does in fact occur between the UO_3 and the N_2O_5 vapor, but admixture of the powders is a useful method of maintaining a high concentration of the pentoxide at the surface of the UO_3 particles.

When the N_2O_5 adduct is heated at 125–130°C. and at 10^{-4} mm. pressure for about 200 min., pure uranyl nitrate is obtained. It is of particular interest that the adduct $UO_2(NO_3)_2 \cdot N_2O_5$ has a lower thermal stability than the corresponding adduct $UO_2(NO_3)_2 \cdot N_2O_4$, which is not fully dissociated under these conditions (Sec. II.B).

B. Dinitrogen Pentoxide in Pure Nitric Acid

Attempts to prepare anhydrous nitrates using 100% nitric acid alone have met with little success, but a solution of dinitrogen pentoxide in pure nitric acid is sometimes found to be an efficient and convenient medium.

1. Thorium(IV) Nitrate

A weighed amount (1–2.5 g.) of $Th(NO_3)_4 \cdot 4H_2O$ is added to 10–20 ml. of anhydrous nitric acid at 25°C., and dinitrogen pentoxide is condensed on to the mixture. When sufficient pentoxide has been condensed, the mixture is distilled under vacuum to remove the nitric acid and unreacted pentoxide. The initial addition of an appropriate excess of dinitrogen pentoxide to the reaction mixture is an essential feature of the preparation. If the number of moles of N_2O_5 beyond that required to react with the water of the tetrahydrate is less than 5.5, impure products containing dinitrogen tetroxide and some water are obtained. Reactions employing 5.5–8.2 moles of excess N_2O_5 give a white granular solid of composition $Th(NO_3)_4 \cdot 2N_2O_5$ (52). This compound is heated at 150–160°C. at 10^{-5} mm. pressure for 4–5 hr., when the N_2O_5 is completely removed and the pure $Th(NO_3)_4$ remains as a white powder.

IV. COMPARISON OF DINITROGEN TETROXIDE AND DINITROGEN PENTOXIDE AS REAGENTS

A comprehensive review of the known anhydrous metal nitrates has been given elsewhere (1). From this, it is clear that in many cases the same products are obtained using dinitrogen pentoxide or dinitrogen tetroxide. In some instances, however, dinitrogen pentoxide offers distinct advantages. This applies particularly where the metal has a range of available oxidation states, and will readily form oxide-nitrates; dinitrogen pentoxide has, therefore, been found especially useful in the preparation of anhydrous nitrates of the early transition metals. Where the metal may form a nitrate or an oxide-nitrate, the former is often more readily obtained by reactions involving the pentoxide. For example, titanium or zirconium halides react with dinitrogen tetroxide to give the oxide-nitrates $TiO(NO_3)_2$ and $ZrO(NO_3)_2$ (54), but by the use of dinitrogen pentoxide the nitrates $Ti(NO_3)_4$ and $Zr(NO_3)_4$ are obtained, as described above. In cases where more than one oxide-nitrate is possible, the more highly nitrated product is usually formed by reactions with the pentoxide; this is illustrated below by reference to vanadium oxide nitrates.

The relative stabilities of the addition compounds formed between metal nitrates and dinitrogen tetroxide or dinitrogen pentoxide also has an important bearing on possible preparative methods. In general, the pentoxide addition compounds are weaker than those formed by the tetroxide, so that the anhydrous nitrates are more readily available by thermal decomposition of the pentoxide addition compounds. This has already been referred to (Sec. III.A) in connection with the uranyl nitrate adducts, and recent work on chromium(III) nitrate provides an excellent example of the wide difference between the complexing powers of the two oxides of nitrogen (55,56).

A. Chromium Compounds

Chromium carbonyl will react with both dinitrogen pentoxide and dinitrogen tetroxide. Even in the presence of excess of the former, chromium(III) nitrate separates in a pure form containing no combined pentoxide (56). Using the tetroxide, the addition compound $Cr(NO_3)_3 \cdot 2N_2O_4$ is invariably obtained, and this compound is so stable that no method has yet been found for removal of the combined tetroxide without simultaneous decomposition of the metal nitrate (55).

1. Chromium(III) Nitrate

The dinitrogen pentoxide evolved from a mixture of phosphorus pentoxide and fuming nitric acid is treated with ozonized oxygen to oxidize the tetroxide impurity. The gas stream is passed into 100 ml. dry carbon tetrachloride to give a practically colorless solution containing 3.0–3.5 g. of dinitrogen pentoxide. A solution of chromium carbonyl (1 g.) in carbon tetrachloride (100 ml.) is added slowly, with stirring, at room temperature. An immediate reaction occurs; gases are evolved, and a green suspension is formed in the solution. The vessel is fitted with a phosphorus pentoxide guard tube and allowed to stand for 12 hr., when the suspension coagulates and settles. It is then filtered using a filter tube protected from atmosphere and washed with carbon tetrachloride. When carbon tetrachloride is removed under vacuum, a pale-green powder remains having the exact composition $Cr(NO_3)_3$. The equation

$$Cr(CO)_6 + 3N_2O_5 \rightarrow Cr(NO_3)_3 + 6CO + 3NO_2$$

involves a pentoxide:carbonyl mole ratio of 3:1, whereas in the above preparation the mole ratio used is 6:1. In spite of this excess of pentoxide there is no indication that a chromium(III) nitrate–dinitrogen pentoxide adduct is formed at any stage in the reaction. The final drying of the precipitate involves vacuum treatment for a short time only and at room temperature, so that the stability of any adduct present at this stage must be very small indeed.

2. Chromium(III) Nitrate Bisdinitrogen Tetroxide

Chromium carbonyl is finely ground and added to excess of liquid dinitrogen tetroxide at room temperature. The carbonyl slowly dissolves, and a grey powder is precipitated from the solution. Reaction is slow, and the reaction mixture is, therefore, allowed to stand for several days. Dinitrogen tetroxide is then removed under vacuum, leaving a free-flowing grey powder which has the composition $Cr(NO_3)_3 \cdot 2N_2O_4$.

This compound begins to decompose at an appreciable rate at 40°C., and studies of its decomposition by chemical analysis, vapor pressure measurement, and thermogravimetric analysis (55) have given no indication that the simple nitrate is a thermal decomposition product. Chromium(III) nitrate itself is unstable above 50°C. (56), so that attempts to isolate this compound from the dinitrogen tetroxide adduct are necessarily restricted to low temperatures; the stability of the adduct is so high under these conditions that no efficient method has yet been found by which the tetroxide can be removed.

B. Vanadium Compounds

Vanadium nitrate, $V(NO_3)_5$, is unknown, but two oxide nitrates $VO(NO_3)_3$ and $VO_2(NO_3)$ have been isolated. Their methods of preparation show how the product obtained is determined by the particular oxide of nitrogen which is used.

1. Vanadium Oxide Trinitrate

Dinitrogen pentoxide reacts with vanadium pentoxide,

$$V_2O_5 + 3N_2O_5 \rightarrow 2VO(NO_3)_3$$

The oxide trinitrate is yellow, melts at 2°C. and boils at 68–70°C. in high vacuum (11,57). Experimental details for this preparation have not been published. Reaction with $VOCl_3$ gives the same product.

2. Vanadium Dioxide Mononitrate

This is the product of reaction in systems containing dinitrogen tetroxide:

$$V + 2N_2O_4 \text{ (in MeCN)} \rightarrow VO_2(NO_3) + 3NO$$

Chips of electrolytic vanadium metal (5 g.) are added to 75–100 ml. dinitrogen tetroxide in a flask cooled to 0–5°C. On adding 5–8 ml. acetonitrile, reaction occurs and a precipitate forms. The reaction mixture should be allowed to stand for 5–6 hr., after which dinitrogen tetroxide and acetonitrile are removed under vacuum. The product is a brick-red powder of composition $VO_2(NO_3)$ (58).

V. HALOGEN NITRATES

Although details of the experiments using halogen nitrates in the preparation of anhydrous metal nitrates have not been published, this field of chemistry has been reviewed by Schmeisser and Brändle (59), and the method may have particular value in the case of those nitrates which exist only at very low temperatures. As more metal nitrates are prepared, it is becoming increasingly evident that the temperature ranges for stability may vary very widely, depending on such factors as the electronic structure and available valency range of the metal, the number of nitrate groups involved, and the readiness with which the metal forms M–O–M polymers. This is illustrated by decomposition temperatures for the nitrates of the alkali metals (475–600°C.); calcium, strontium, and barium (575–675°C.); cobalt, nickel, copper, and zinc (270–350°C.); beryllium (125°C.); and chromium (50°C.). It appears likely that metal nitrates which are at present unknown may become available if reactions can be carried out at sufficiently low temperatures.

The halogen nitrates, and particularly chlorine nitrate, have distinct advantages in this respect. Chlorine nitrate is prepared when chlorine monoxide and dinitrogen pentoxide are condensed together in a liquid-air bath:

$$Cl_2O + N_2O_5 \rightarrow 2ClNO_3$$

The mixture is held for some hours at $-80\,^{\circ}$C. to complete the reaction (60). The compound may also be prepared by reaction of chlorine monoxide with dinitrogen tetroxide (61)

$$2Cl_2O + N_2O_4 \rightarrow 2ClNO_3 + Cl_2$$

Chlorine nitrate is a yellow liquid, melting point $-107\,^{\circ}$C., boiling point (by extrapolation) $+18\,^{\circ}$C. (62). The chlorine atom may be considered to carry a partial positive charge $Cl^{\delta+} - ONO_2^{\delta-}$, so that in its reactions the compound is a ready source of nitrate ions (11,59). A further possible advantage in the use of chlorine nitrate lies in the fact that its reactions are not complicated by the NO^+ or NO_2^+ ions which are inevitably present in reactions of N_2O_4 or N_2O_5.

Because of its low melting point, reactions with chlorine nitrate, e.g.,

$$TiCl_4 + 4ClNO_3 \rightarrow 4Cl_2 + Ti(NO_3)_4$$

can be carried out conveniently at the temperature of solid carbon dioxide ($-80\,^{\circ}$C.) and chlorine, with excess chlorine nitrate, can be removed in vacuum at this temperature (11,59). The nitrates $B(NO_3)_3$ ($-78\,^{\circ}$C.), $Al(NO_3)_3$ ($-7\,^{\circ}$C.), and $Sn(NO_3)_4$ ($-60\,^{\circ}$C.) are said to be prepared in this way at the temperatures given.

Bromine trinitrate prepared by the reactions

$$BrF_3 + 3N_2O_5 \rightarrow Br(NO_3)_3 + 3NO_2F$$

or

$$BrF_3 + 3HNO_3 \rightarrow Br(NO_3)_3 + 3HF$$

has been employed in the preparation of tin tetranitrate. However, it melts with decomposition at $48\,^{\circ}$C., and does not possess the physical advantages of chlorine nitrate (11,59).

REFERENCES

1. Addison, C. C., and N. Logan, *Advan. Inorg. Chem. Radiochem.* 6, to be published.
2. Wallwork, S. C., *Proc. Chem. Soc.* 1959, 311; *J. Chem. Soc.* to be published; B. Duffin, and S. C. Wallwork, *Acta Cryst.* 16, A33 (1963).
3. Addison, C. C., M. Kilner, and A. Wojcicki, *J. Chem. Soc.* 1961, 4839.
4. Clark, H. C., and R. J. O'Brien, *Inorg. Chem.* 2, 740 (1963).
5. Addison, C. C., and M. Kilner, unpublished results.
6. Hoard, J. L., and J. D. Stroupe, Atomic Energy Project Report 1943, A1229; W. H. Zachariasen, *Acta Cryst.* 7, 795 (1954).
7. Gatehouse, B. M., and A. E. Comyns, *J. Chem. Soc.* 1958, 3965. J. G. Allpress and A. N. Hambly, *Australian J. Chem.* 12, 569 (1959). V. M. Vdovenko, C. V. Stroganov, and A. P. Sokolov, *Radiokhimiya* 1, 97 (1963) and refs. therein.
8. Straub, D. K., R. S. Drago, and J. T. Donoghue, *Inorg. Chem.* 1, 848 (1962).
9. Addison, C. C., and A. Walker, *Proc. Chem. Soc.* 1961, 242.

10. Addison, C. C., and B. J. Hathaway, *J. Chem. Soc.* **1958**, 3099.
11. Schmeisser, M., *Angew. Chem.* **67**, 493 (1955); M. Schmeisser and K. Brändle, *Angew. Chem.* **69**, 781 (1957).
12. Field, B. O., and C. J. Hardy, *Proc. Chem. Soc.* **1962**, 76.
13. Addison, C. C., B. J. Hathaway, N. Logan, and A. Walker, *J. Chem. Soc.* **1960**, 4308.
14. Addison, C. C., and A. Walker, *J. Chem. Soc.* **1963**, 1220.
15. Addison, C. C., *Advan. Chem. Ser.* **36**, 131 (1962).
16. Field, B. O., and C. J. Hardy, *J. Chem. Soc.* **1963**, 5278.
17. Naumann, A., *Chem. Ber.* **37**, 4333 (1904).
18. Guntz, A., and M. Martin, *Bull. Soc. Chim. France* **7**, 313 (1910).
19. Hathaway, B. J., and A. E. Underhill, *J. Chem. Soc.* **1960**, 648.
20. Jander, G., and H. Wendt, *Z. Anorg. Allgem. Chem.* **258**, 1 (1949).
21. Jander, G., and K. Brodersen, *Z. Anorg. Allgem. Chem.* **262**, 33 (1950).
22. Audrieth, L. F., E. E. Jukkola, R. E. Meints, and B. S. Hopkins, *J. Am. Chem. Soc.* **53**, 1805 (1931).
23. Addison, C. C., J. Allen, H. C. Bolton, and J. Lewis, *J. Chem. Soc.* **1951**, 1289.
24. Addison, C. C., and R. Thompson, *J. Chem. Soc.* **1949**, S.218.
25. Clusius, K., and M. Vecchi, *Helv. Chim. Acta* **36**, 930 (1953).
26. Millen, D. J., and D. Watson, *J. Chem. Soc.* **1957**, 1369.
27. Goulden, J. D. S., and D. J. Millen, *J. Chem. Soc.* **1950**, 2620.
28. Moeller, T., and V. D. Aftandilian, *J. Am. Chem. Soc.* **76**, 5249 (1954); T. Moeller, V. D. Aftandilian, and G. W. Cullen, in *Inorganic Syntheses*, Vol. 5, T. Moeller, ed., McGraw-Hill, New York, 1957, p. 37.
29. Addison, C. C., and R. Thompson, *Nature* **162**, 369 (1948).
30. Addison, C. C., and R. Thompson, *J. Chem. Soc.* **1949**, S.211.
31. Addison, C. C., J. Lewis, and R. Thompson, *J. Chem. Soc.* **1951**, 2829.
32. Addison, C. C., and H. A. J. Champ, unpublished results.
33. Addison, C. C., J. Lewis, and R. Thompson, *J. Chem. Soc.* **1951**, 2838.
34. Addison, C. C., N. Hodge, and J. Lewis, *J. Chem. Soc.* **1953**, 2631.
35. Addison, C. C., J. C. Sheldon, and N. Hodge, *J. Chem. Soc.* **1956**, 3900.
36. Addison, C. C., and N. Hodge, *J. Chem. Soc.* **1964**, to be published.
37. Gibson, G., and J. J. Katz, *J. Am. Chem. Soc.* **73**, 5436 (1951).
38. Jezowska-Trzebiatowska, B., and B. Kedzia, *Bull. Acad. Polon. Sci., Ser. Sci. Chim.* **10**, 213 (1962).
39. Addison, C. C., and N. Hodge, *J. Chem. Soc.* **1961**, 2490.
40. Addison, C. C., and J. C. Sheldon, *J. Chem. Soc.* **1958**, 3142.
41. Addison, C. C., and J. C. Sheldon, *J. Chem. Soc.* **1956**, 1941.
42. Addison, C. C., and J. C. Sheldon, *J. Chem. Soc.* **1956**, 2709.
43. Rubin, B., H. H. Sisler, and H. Schechter, *J. Am. Chem. Soc.* **74**, 877 (1952).
44. Addison, C. C., C. P. Conduit, and R. Thompson, *J. Chem. Soc.* **1951**, 1303.
45. Addison, C. C., and J. C. Sheldon, *J. Chem. Soc.* **1956**, 2705.
46. Addison, C. C., and N. Logan, unpublished results.
47. Addison, C. C., B. J. Hathaway, and D. Sutton, unpublished results.
48. Addison, C. C., and B. F. G. Johnson, unpublished results.
49. Addison, C. C., and B. C. Smith, unpublished results.
50. Straub, D. K., H. H. Sisler, and G. E. Ryschkewitsch, *J. Inorg. Nucl. Chem.* **24**, 919 (1962).
51. Addison, C. C., and B. J. Hathaway, unpublished results.
52. Ferraro, J. R., L. I. Katzin, and G. Gibson, *J. Am. Chem. Soc.* **77**, 327 (1955).
53. Gibson, G., C. D. Beintema, and J. J. Katz, *J. Inorg. Nucl. Chem.* **15**, 110 (1960).

54. Gutmann, V., and H. Tannenberger, *Monatsh. Chem.* **87**, 421 (1956).
55. Addison, C. C., and A. H. Norbury, unpublished results.
56. Addison, C. C., and D. J. Chapman, *J. Chem. Soc.* **1964**, 539.
57. Schmeisser, M., and D. Lützow, *Angew. Chem.* **66**, 230 (1954).
58. Panontin, J. A., A. K. Fischer, and E. A. Heintz, *J. Inorg. Nucl. Chem.* **14**, 145 (1960).
59. Schmeisser, M., and K. Brändle, *Angew. Chem.* **73**, 388 (1961).
60. Schmeisser, M., W. Fink, and K. Brändle, *Angew. Chem.* **69**, 780 (1957).
61. Martin, H., and Th. Jacobsen, *Angew. Chem.* **67**, 524 (1955).
62. Martin, H., *Angew. Chem.* **70**, 97 (1958).

CHAPTER 7

Halogen and Halogenoid Derivatives of the Silanes

ALAN G. MACDIARMID

University of Pennsylvania, Philadelphia, Pennsylvania

CONTENTS

I. INTRODUCTION

The silanes may be regarded as the silicon analogs of the alkanes, and they form an homologous series of compounds of general formula

Si_nH_{2n+2}. Although all the compounds in the series from $n = 1$ to $n = 8$ have been isolated, only silane, SiH_4, disilane, Si_2H_6, and trisilane, Si_3H_8, have been thoroughly characterized (1–4). Isomers of some of the higher silanes have also been identified (5).

A large number of halogen and halogenoid derivatives of silane are known, and more recently several monohalodisilanes have been synthesized. The mono-halo derivatives of silane and disilane are of particular importance because they serve as starting materials for the preparation of many derivatives containing the silyl, —SiH_3, group or the disilanyl, —SiH_2SiH_3, group, respectively. These species are of considerable interest since silyl compounds may be regarded as the silicon analogs of methyl compounds, and similarly disilanyl compounds may be considered to be the silicon analogs of ethyl compounds. Silyl and disilanyl compounds not discussed in this chapter are described in recent monographs (1,1a) and review articles (2,3).

Many organic and inorganic silicon compounds may be regarded as derivatives of the halosilanes. The compound SiH_3I—iodosilane or silyl iodide—for example, may be considered to be the parent member of the series CH_3SiH_2I, $(CH_3)_2SiHI$, and $(CH_3)_3SiI$ where one, two, or three hydrogen atoms, respectively, are replaced by methyl groups. Similarly, if the hydrogen atoms are replaced by a halogen such as chlorine to give compounds such as $ClSiH_2I$, Cl_2SiHI, and Cl_3SiI, these substances can also be regarded as derivatives of SiH_3I. A similar situation applies in the case of $(C_2H_5O)_2SiHCl$ and related compounds. It is, therefore, obvious that in a chapter of this size it is impossible to attempt to describe the enormous number of compounds which may be classified as halogen or halogenoid derivatives of the silanes. The following discussion will consequently be limited to inorganic compounds, although in certain instances it will be appropriate or desirable to refer to organosilicon compounds.* The only compounds which will be described are silicon–halogen or silicon–halogenoid species which may contain, in addition to the above linkages, one or more silicon–hydrogen or silicon–silicon bonds. Hence, the relatively large group of compounds which contain other elements attached to silicon such as $(SiBr_3)_2O$ (12), [Si-$(NCO)_3]_2O$ (13), $(SiCl_3)_3N$ (14), $(SiCl_3)_2S$ (15), etc., will not be discussed.

All the compounds to be described hydrolyze readily, even in moist air, and some, such as the disilanyl halides, are spontaneously inflammable in air. For these reasons, and also because for many years the preparation of compounds such as SiH_3I involved the use of silane, which is spon-

---

* For further information on inorganic or organic silicon compounds, the reader is referred to several books or review articles (1–4, 6–11) which describe various aspects of this large field of chemistry.

taneously inflammable in air, vacuum-system techniques have been employed in the synthesis of many of these compounds. Until very recently, the preparation of relatively large quantities of, for example, SiH₃I has been a most tedious process. If this were to be converted to some other silyl derivative, then the use of vacuum-system techniques was almost essential, since the quantities of materials required in order to study thoroughly a reaction in the vacuum system is generally very much less than that needed for even small-scale bench experiments. However, SiH₃I can now be made in relatively large quantities and, hence, it should now be possible to use it in certain reactions employing standard bench techniques. Some derivatives of the silicon hydrides containing silicon–hydrogen bonds, such as SiHCl₃, can be synthesized relatively easily in large quantities (16,17) or purchased. Hence, although vacuum-system techniques are required in many operations involving derivatives of the silanes, considerable work may be done using standard chemical preparative methods. It seems very likely that some syntheses, which have been carried out in the past with vacuum-system techniques because of scarcity of materials, might be adapted to standard bench techniques in view of the present greater availability of certain compounds. However, considerable care should be exercised when performing experiments on the bench which have previously only been carried out in a vacuum system. Particularly, in the case of compounds containing an —SiH₃ group, small quantities of silane may be evolved which could cause serious explosions.

Although several books are available (18–23) which describe the basic techniques for handling volatile materials and performing reactions in vacuum systems, it is a somewhat difficult matter for a person to learn these techniques unless he is given personal instruction. It is therefore not advisable to handle substances such as silane or disilane which can explode violently in a vacuum system in the presence of even small quantities of air, until the operator has become completely familiar and experienced with the necessary manipulative procedures involved. It is also highly desirable to use liquid nitrogen as a refrigerant rather than liquid air or liquid oxygen when handling substances which are combustible. If a trap should break when immersed in either of the latter two materials, the possibility of an explosion occurring is greatly increased.

II. GENERAL METHODS OF SYNTHESIS

There are several general types of reactions by which halogen or halogenoid derivatives of the silanes may be synthesized. These are discussed briefly in this section, and their application to the synthesis of

specific silicon compounds is described, together with less general types
of reactions, in the following sections.

It should be held in mind that all halogen and halogenoid derivatives
of the silanes are readily hydrolyzed even by atmospheric moisture and
appropriate precautions to guard against hydrolysis must, therefore, always be taken. Some compounds containing silicon–silicon or silicon–
hydrogen bonds in the molecule are also spontaneously inflammable in
air. Those which are not spontaneously inflammable may also contain,
or decompose to yield, substances which are spontaneously inflammable,
and it is advisable to handle them with care. The inflammability of
many compounds which have been prepared only by vacuum-system
techniques has not been examined. It is therefore wise to consider
them to be spontaneously inflammable, until it is proven otherwise.

A. From Silicon or Silicides

On heating, the free halogens all react with silicon or metal silicides
to form silicon tetrahalides and in some cases the halides of higher silanes
are also formed.

Volatile compounds containing silicon–hydrogen bonds may be formed
very readily in somewhat analogous reactions by heating silicon or a
metal silicide in a current of the appropriate dry gaseous hydrogen halide.
The trihalosilanes, $SiHCl_3$ (16,17,24), $SiHBr_3$ (25,26), and $SiHI_3$ (27)
have been obtained in good yields by this method. The tetrahalides
and halides of the higher silanes may also be formed in these reactions.
Silicides such as ferrosilicon (28), copper silicide (1,29), etc., have been
used. If the hydrogen halide is mixed with hydrogen in reactions of the
above types, then dihalosilanes may be formed in addition to the trihalosilanes (1). On heating, silicon also reacts with a number of organic
halides in the presence of certain catalysts such as metallic copper to
give organohalosilanes such as $(CH_3)_2SiCl_2$ (9,30) or CH_3SiHCl_2 (31),
etc. This is the basis for the industrial preparation of organohalosilanes
which are used in large quantities in the preparation of silicone polymers.

B. From Reactions of the Silicon–Silicon Bond

The cleavage of a silicon–silicon bond in a di- or tri-silane has been
used in a few cases to form a silicon derivative containing a specific halogen. For example, when Si_2F_6 is heated with chlorine, F_3SiCl is formed
(32). The reaction is complicated by the simultaneous formation of

$$Si_2F_6 + Cl_2 \rightarrow 2F_3SiCl \qquad\qquad 1$$

other compounds such as SiF_2Cl_2. The chief disadvantage of the method is that compounds containing silicon–silicon bonds are usually less readily available than derivatives of silane. Also, the method is limited to compounds which do not contain silicon–hydrogen bonds since these bonds will be halogenated simultaneously with the silicon–silicon bonds.

C. From Reactions of the Silicon–Hydrogen Bond

A hydrogen atom attached to silicon may be replaced directly by a halogen by reacting the silicon hydride with free halogen or with a gaseous hydrogen halide in the presence of the appropriate aluminum halide catalyst. These types of reactions have been extensively investigated— particularly the latter—for all halogens except fluorine.

Reaction of the free halogen with a silicon–hydrogen bond is, in general, much more violent and less controllable than the corresponding reaction with hydrogen halide and aluminum halide (4). As expected, the violence of reaction decreases from chlorine to bromine to iodine. The rate of the reaction also tends to decrease as the number of silicon– hydrogen bonds in the molecule decreases (2,33). The reaction of chlorine or bromine with SiH_4 is explosively violent at room temperature (34), but at low temperatures the reactions can be made to occur more slowly and controllably. For example, at $-80\,°C.$ solid bromine reacts to give SiH_3Br and some SiH_2Br_2 (34):

$$SiH_4 + Br_2 \rightarrow SiH_3Br + HBr \qquad\qquad 2$$

Partial replacement of hydrogen by halogens is possible in the case of organosilicon hydrides by carrying out the reaction in an inert solvent (33):

$$C_6H_5SiH_3 + Br_2 \rightarrow C_6H_5SiH_2Br + HBr \qquad\qquad 3$$

The most convenient method for controllably replacing silicon–hydrogen bonds in SiH_4 with a halogen involves the reaction of SiH_4 with gaseous hydrogen halide in the presence of the appropriate anhydrous aluminum halide catalyst (2). It has been shown with SiH_4 that reaction will not occur, even on heating, unless the catalyst is present (35). In general, reaction times of from 6 to 24 hr. at $100\,°C.$ at atmospheric pressure or below, give good yields of the desired product. By controlling the proportions of SiH_4 and hydrogen halide employed, any of the substitution products SiH_3X, SiH_2X_2, $SiHX_3$, or SiX_4 desired, can be obtained in good yield although smaller quantities of the other substituted products will always be present (35,36):

$$SiH_4 + HCl \xrightarrow{Al_2Cl_6} SiH_3Cl + H_2 \qquad\qquad 4$$

A similar type of reaction occurs with organosilicon hydrides. In this way, CH_3SiH_3, for example, may be converted to CH_3SiH_2Cl (35). In some cases, however, hydrogen halides will react with silicon–hydrogen bonds in the absence of aluminum halide catalyst. Thus at 120 °C., hydrogen iodide and $C_6H_5SiH_3$ give $C_6H_5SiH_2I$ in addition to other products (37). As shown in Section II.G, better methods of synthesis of silyl halides than those involving the use of dangerously explosive silane described above are now available. However, at the present time this method is the only one known for preparing monohalogen derivatives of disilane:

$$Si_2H_6 + HI \xrightarrow{Al_2I_6} Si_2H_5I + H_2 \qquad\qquad 5$$

It should be noted that disilane undergoes reaction with HBr, HCl, and HI (38,39) at lower temperatures and at a greater rate than silane.

A very convenient method for directly replacing the hydrogen of a silicon–hydrogen bond with a halogen or halogenoid involves the reduction of a metal halide or halogenoid such as $HgCl_2$, $AgCl$, $CuCl_2$, $AgNCO$, etc., by a silicon–hydrogen bond (40–42). The metal salt is added gradually to an excess of the pure silane at room temperature or at reflux temperature or to a boiling solution of the silane in an inert solvent such as carbon tetrachloride:

$$n\text{-}C_7H_{15}SiH_3 + HgBr_2 \rightarrow n\text{-}C_7H_{15}SiH_2Br + HBr + Hg \qquad 6$$

$$cyclo\text{-}C_6H_{11}SiH_3 + 2AgNCO \rightarrow cyclo\text{-}C_6H_{11}SiH_2NCO + 2Ag + HNCO \qquad 7$$

More than one hydrogen atom may be replaced by this method if a greater proportion of the metal salt is used.

In a somewhat analogous type of reaction, certain organic halides will convert a silicon–hydrogen bond to a silicon–halogen bond. For example, $(C_2H_5)_3SiH$ reacts with $n\text{-}C_6H_{13}Cl$ in the presence of aluminum chloride catalyst at reflux temperature as shown below (24):

$$(C_2H_5)_3SiH + n\text{-}C_6H_{13}Cl \xrightarrow{Al_2Cl_6} (C_2H_5)_3SiCl + n\text{-}C_6H_{14} \qquad 8$$

Similar reactions are also observed with the higher silanes. For example, Si_3H_8 and chloroform react in the presence of aluminum chloride catalyst at 50–70 °C. (43):

$$Si_3H_8 + 4CHCl_3 \rightarrow Si_3H_4Cl_4 + 4CH_2Cl_2 \qquad 9$$

It should be noted that organohalosilanes such as $(C_2H_5)_2SiHCl$ have been obtained by mixing a Grignard reagent with excess $SiHCl_3$ (44).

Redistribution reactions involving silicon–hydrogen bonds are discussed in Section II.H.

D. From Reactions of the Silicon–Halogen Bond

The ready replacement of a halogen in a silicon–halogen bond by a different halogen or by a halogenoid, or by some other atom or group, is one of the most useful reactions in silicon chemistry. Such a replacement can usually be accomplished by permitting the halosilane in the gaseous or liquid state to come in contact with an appropriate metal halide or halogenoid (2). Mild heating, particularly in the case of organosilicon halides is sometimes necessary (45–47).

A "Conversion Series" has been drawn up for the reaction of silyl compounds with silver salts which indicates how certain bonds attached to silicon may be interconverted. Treatment of a compound with the appropriate silver salt will bring about a conversion into any compound later in the series but into none earlier therein (3,45–48,48a). It is based on a very similar series formulated earlier for triorganosilyl compounds (47), and all parent silyl compounds so far investigated substantiate the order:

$$SiH_3I \rightarrow (SiH_3)_2Se \rightarrow (SiH_3)_2S \rightarrow SiH_3Br \rightarrow SiH_3NCSe \rightarrow$$
$$(SiH_3N)_2C \rightarrow SiH_3Cl \rightarrow SiH_3CN \rightarrow SiH_3NCS \rightarrow SiH_3NCO \rightarrow$$
$$(SiH_3)_2O \rightarrow SiH_3F$$

For example, SiH_3I reacts with AgCN to give SiH_3CN (49)

$$SiH_3I + AgCN \rightarrow SiH_3CN + AgI \qquad\qquad 10$$

but $(SiH_3)_2O$ does not react with AgCl to give SiH_3Cl. There is also evidence for a similar conversion series involving mercuric salts but only a few reactions have been investigated so far (3). Certain conversions in the case of organosilicon compounds have also been found to occur with mercuric, plumbic, cupric, and cadmium salts (47). It is also highly probable that these, and other metal salts, would bring about similar conversions with inorganic silicon compounds. Although the above series is postulated for compounds containing the —SiH_3 group, it appears likely that it applies equally well to di-, tri-, and tetrahalosilanes as well as to disilanyl halides. In fact, it has been observed that SiH_2I_2 may be converted to SiH_2Cl_2 in fairly good yields by passing the vapors of the iodide over silver chloride at room temperature (50). Disilanyl iodide, Si_2H_5I, has been converted in a similar manner to Si_2H_5Cl (51).

Silyl iodide is the most useful silyl halide in the Conversion Series since it appears that it should be possible to convert it to all the other compounds in the series. Indeed, it has already been used for the preparation of $(SiH_3)_2Se$ (52), SiH_3CN (49), SiH_3Cl (53), SiH_3NCS (49), SiH_3NCO (48), SiH_3NCSe (48a), $(SiH_3N)_2C$ (48a), and $(SiH_3)_2S$ (52) by reaction with the appropriate silver salts. This is most fortunate since

SiH_3I can now be readily synthesized as shown in Section III.D without the use of silane, which is somewhat dangerous to handle in large quantities.

It should be noted that when the silicon compound contains a silicon–hydrogen bond, the use of silver fluoride to prepare the silicon fluoride is not satisfactory since oxidation of the silicon–hydrogen bond occurs (16). In this case it is desirable to use fluorides such as SbF_3, SnF_4, and TiF_4 (16,54,55). By using SbF_3 (with $SbCl_5$ catalyst), good yields of SiH_3F, SiH_2F_2, and $SiHF_3$ are obtained from the corresponding chlorides (16):

$$3SiH_2Cl_2 + 2SbF_3 \rightarrow 3SiH_2F_2 + 2SbCl_3 \qquad\qquad \textbf{11}$$

However, as shown in Section III.A, only the SiH_2F_2 obtained by this method appears to be completely pure and pure SiH_3F is better obtained by other methods. Antimony trifluoride is also useful in partially replacing halogens in silicon tetrahalides. For example, $SiCl_4$ reacts exothermically with a deficit of SbF_3 in the presence of $SbCl_5$ catalyst to give a mixture from which $SiFCl_3$, SiF_2Cl_2, SiF_3Cl, and SiF_4 can be obtained (56).

In the case of azides it has been possible to replace a Si–Cl bond with a Si–N_3 bond merely by refluxing a solution of the silicon chloride with sodium azide in the presence of catalytic amounts of $LiAlH_4$ (57) or aluminum azide (58), e.g.,

$$(CH_3)_3SiCl + NaN_3 \rightarrow (CH_3)_3SiN_3 + NaCl \qquad\qquad \textbf{12}$$

Although attempts have been made to partially reduce silicon halides by means of hydride reducing agents such as $LiAlH_4$, complete reduction always takes place. Thus with $SiCl_4$, only SiH_4 is obtained and no species such as SiH_3Cl have been isolated. Partial reduction has been accomplished however, in some cases, by passing the vapor of a chlorosilane mixed with hydrogen over heated aluminum or zinc at 400°C. (59). In this manner, for example, $SiCl_4$ has been converted in 25% yield to $SiHCl_3$ and CH_3SiCl_3 has been converted to CH_3SiHCl_2:

$$6SiCl_4 + 3H_2 + 2Al \rightarrow 6SiHCl_3 + Al_2Cl_6 \qquad\qquad \textbf{13}$$

These reactions may involve the intermediate formation of a metal hydride which then undergoes metathesis with the silicon halide, but no evidence for such a mechanism has been obtained.

Redistribution reactions involving silicon–halogen bonds are discussed in Section II.H.

E. From Reactions of the Silicon–Oxygen Bond

Compounds containing silicon–oxygen bonds frequently undergo interesting reactions with certain anhydrous hydrogen halides (and hydrofluoric acid) and other covalent halides to form halosilanes.   In a few instances, iodine has been observed to react in a similar manner.

Cleavage of a silicon–oxygen bond by hydrofluoric acid is illustrated, for example, by the formation of cyclo-$C_6H_{11}SiH_2F$ upon heating (cyclo-$C_6H_{11}SiH_2)_2O$ with 48% aqueous HF for 10 min. at 60°C. (42):

$$(\text{cyclo-}C_6H_{11}SiH_2)_2O + 2HF \;\rightarrow\; 2\,\text{cyclo-}C_6H_{11}SiH_2F + H_2O \qquad\qquad \textbf{14}$$

This is completely analogous to the well-known reaction

$$SiO_2 + 4HF \;\rightleftarrows\; SiF_4 + 2H_2O \qquad\qquad \textbf{15}$$

The driving force behind this reaction depends greatly upon the formation of the very strong Si–F bonds (60) (Si–O $=$ 108 kcal./mole; Si–F $=$ 135 kcal./mole); consequently, the analogous reaction with hydrogen chloride (Si–Cl $=$ 91 kcal./mole) does not proceed so readily although 25% yields of cyclo-$C_6H_{11}SiH_2Cl$ can be obtained if hot hydrogen chloride is used in the presence of P_2O_5 (to remove water and force the equilibrium to the right) (42).   There is some evidence that hydrogen iodide might be capable of cleaving the silicon–oxygen bond in $(CH_3SiH_2)_2O$ to give CH_3SiH_2I (61).   Triorganosilyl fluorides (and chlorides) are frequently formed from the corresponding disiloxane* and hydrogen fluoride (and chloride) generated *in situ* by warming the disiloxane with a mixture of concentrated sulfuric acid and appropriate ammonium halide (62).

It should be noted that hydrofluoric acid may also be used to replace chlorine atoms directly with fluorine in certain silicon compounds. Thus warm 48% hydrofluoric acid readily converts cyclo-$C_6H_{11}SiCl_3$ to cyclo-$C_6H_{11}SiF_3$ in a one step reaction (42).   This type of reaction may be regarded as proceeding through hydrolysis of the chloride to give the intermediate siloxane (cyclo-$C_6H_{11}SiO_{1.5})_x$, in which the silicon–oxygen bonds are then cleaved by the hydrofluoric acid.

The silicon–oxygen bond is cleaved by some covalent or semicovalent halides either at low temperatures or upon gentle heating (62).   For example, at -78°C. $(SiH_3)_2O$ and BF_3 react smoothly to give SiH_3F (63):

$$(SiH_3)_2O + BF_3 \;\rightarrow\; SiH_3OBF_2 + SiH_3F \qquad\qquad \textbf{16}$$

and the SiH_3OBF_2 also formed subsequently decomposes slowly at room temperature:

$$3SiH_3OBF_2 \;\rightarrow\; 3SiH_3F + B_2O_3 + BF_3 \qquad\qquad \textbf{17}$$

---

* The term siloxane is the name given to a silicon ether. Thus $(SiH_3)_2O$ is correctly named disiloxane although it is frequently referred to as disilyl ether, especially when it is desirable to stress its relationship to its carbon analog, $(CH_3)_2O$.   The nomenclature used in silicon chemistry is readily available from several sources (6,10,64).

In a similar manner, certain of the boron halides react rapidly at low temperatures with other disiloxanes such as $(CH_3SiH_2)_2O$, $[(CH_3)_2-SiH]_2O$, or $[(CH_3)_3Si]_2O$ to give analogous products (61). Alkoxy silanes may also be cleaved by BF_3. Thus, CH_3OSiH_3 and BF_3 react at $-96°C.$ to yield SiH_3F (65):

$$CH_3OSiH_3 + BF_3 \rightarrow SiH_3F + CH_3OBF_2 \qquad 18$$

The cleavage of a silicon–oxygen bond by BF_3 probably represents the most convenient method for obtaining absolutely pure silicon fluorides such as SiH_3F, CH_3SiH_2F, etc. As expected from steric considerations (2), reaction tends to occur more readily with unsubstituted hydride compounds such as $(SiH_3)_2O$ than with substituted derivatives such as $[(CH_3)_3Si]_2O$. For example, $(SiH_3)_2O$ reacts with Al_2Cl_6 at low temperatures to give SiH_3Cl (66):

$$2(SiH_3)_2O + Al_2Cl_6 \rightarrow 2SiH_3OAlCl_2 + 2SiH_3Cl \qquad 19$$

The SiH_3OAlCl_2 which is also formed decomposes at low temperatures with the liberation of more SiH_3Cl. In the case of $[CH_3(C_2H_5)_2Si]_2O$ rapid reaction only occurs on heating to $135–210°C.$ and the $CH_3(C_2H_5)_2-SiOAlCl_2$ which is formed is stable (67).

It is interesting to note that at room temperatures, iodine has been found to cleave the silicon–oxygen bond in certain disiloxanes to yield a silyl iodide. Thus both $(SiH_3)_2O$ (52) and CH_3OSiH_3 (65) give SiH_3I and $(cyclo-C_6H_{11}SiH_2)_2O$ forms $cyclo-C_6H_{11}SiH_2I$ (42):

$$3(cyclo-C_6H_{11}SiH_2)_2O + 3I_2 \rightarrow$$
$$3\ cyclo-C_6H_{11}SiH_2I + (cyclo-C_6H_{11}SiHO)_3 + 3HI \quad 20$$

Although silyl sulfides and selenides do not appear to have much practical value in any large scale preparation of silyl halides it should be noted that the Si–S and Si–Se bonds in $(SiH_3)_2S$ and $(SiH_3)_2Se$ are cleaved by iodine at room temperature to give high yields of SiH_3I and sulfur or selenium, respectively (52). The compounds also react readily with HI to give SiH_3I and the corresponding hydrogen sulfide or selenide (52).

F. From Reactions of the Silicon–Nitrogen Bond

Silicon amines, in general, react with anhydrous hydrogen halides or halogenoids, or covalent or semicovalent halides, in a manner similar to the siloxanes except that reaction tends to proceed more readily. Silicon–nitrogen bonds are cleaved without heating, often at low temperatures, to yield the corresponding silicon halide. Thus, $(SiH_3)_3N$ (74) reacts instantly with gaseous hydrogen chloride as shown below (68):

$$(SiH_3)_3N + 4HCl \rightarrow 3SiH_3Cl + NH_4Cl \qquad 21$$

This type of reaction appears to offer a convenient method for transforming a silicon halide into a different silicon halide even in a direction reverse to that given by the silver salt Conversion Series. Thus, by using silver chloride it is possible to convert a silicon iodide to a silicon chloride, but it is not possible to convert the chloride back to the iodide by means of silver iodide. However, if, for example, $SiHCl_3$ is first converted to $(SiHN)_x$ or to $SiH(NHC_6H_5)_3$ by means of ammonia or aniline, respectively, then $SiHI_3$ may be readily obtained from these compounds by treating them with dry hydrogen iodide (69):

$$SiHCl_3 + 6C_6H_5NH_2 \rightarrow SiH(NHC_6H_5)_3 + 3C_6H_5NH_2 \cdot HCl \qquad 22$$

$$SiH(NHC_6H_5)_3 + 6HI \rightarrow SiHI_3 + 3C_6H_5NH_2 \cdot HI \qquad 23$$

It might even be possible to replace the HI by a covalent iodide such as PI_3 since reactions of the type

$$4(CH_3)_3SiNHC_6H_5 + GeBr_4 \rightarrow 4(CH_3)_3SiBr + Ge(NHC_6H_5)_4 \qquad 24$$

are known to occur readily (70). It is of interest to note that HN_3 behaves similarly to the hydrogen halides in its reaction with silicon–nitrogen bonds and it has recently been used in the preparation of both SiH_3N_3 (71) and $(CH_3)_3SiN_3$ (72). Hydrogen cyanide also acts in an analogous way with $[(CH_3)_3Si]_2NH$ to give $(CH_3)_3SiCN$ (73).

It seems that the above types of reactions have not been exploited as much as they might be, particularly in the preparation of compounds containing both silicon–hydrogen and silicon–halogen or halogenoid bonds.

Reactions analogous to those observed with the disiloxanes take place between covalent halides and compounds containing silicon–nitrogen bonds. Thus, $(SiH_3)_3N$ reacts at $-78°C$. with both BF_3 and BCl_3 to give the solid addition compounds $(SiH_3)_3N \cdot BF_3$ and $(SiH_3)_3N.BCl_3$, respectively. On warming to room temperature, these compounds decompose smoothly to give $(SiH_3)_2NBF_2$ or $(SiH_3)_2NBCl_2$ and the corresponding silyl halide, e.g.,

$$(SiH_3)_3N \cdot BF_3 \rightarrow (SiH_3)_2NBF_2 + SiH_3F \qquad 25$$

Both $(SiH_3)_2NBF_2$ and $(SiH_3)_2NBCl_2$ decompose on standing at room temperature with the liberation of more silyl halide. In the case of $(SiH_3)_2NBF_2$, it appears that a silyl borazine might be formed (74,75):

$$3(SiH_3)_2NBF_2 \rightarrow (SiH_3NBF)_3 + 3SiH_3F \qquad 26$$

This type of reaction is also an excellent method for obtaining very pure SiH_3F (75). Similar reactions also take place with related compounds such as $(Si_2H_5)_3N$ (76), $(SiH_3)_2NCH_3$, $SiH_3N(CH_3)_2$ (75), $(CH_3SiH_2)_3N$, $CH_3SiH_2N(CH_3)_2$ (77), and $[(CH_3)_3Si]_2NH$ (78,79), etc.

Hexamethyldisilazane, $[(CH_3)_3Si]_2NH$ has been found to undergo a number of interesting reactions with covalent or semicovalent halides such as BF_3, BCl_3, PCl_3, $SbCl_3$, $BiCl_3$, Al_2Cl_6, below room temperature, or upon refluxing, to yield the corresponding trimethylsilyl halide (79). For example, with PCl_3 the following reaction occurs:

$$[(CH_3)_3Si]_2NH + PCl_3 \rightarrow (CH_3)_3SiNHPCl_2 + (CH_3)_3SiCl \qquad 27$$

It may be noted that the Si–P bond in SiH_3PH_2 is cleaved by HBr in an analogous manner to the Si–N bond, to yield SiH_3Br and PH_3 (79a).

G. From Reactions of the Silicon–Phenyl Bond*

The synthesis of silicon halides by the cleavage of a silicon–phenyl bond with anhydrous hydrogen halide is one of the most important recent developments in the inorganic chemistry of silicon. By this method it is not possible, for the first time, to prepare important starting materials such as SiH_3I in relatively large quantities without having to use SiH_4 as an intermediary (37,80–84).

The cleavage of phenylsilane, $C_6H_5SiH_3$, and certain of its derivatives with dry hydrogen halides occurs in the absence of solvent and catalyst at low temperatures. Quantitative yields of halosilanes are obtained in many cases. Some typical reactions are listed below:

$$C_6H_5SiH_3 + HI \xrightarrow{-40°C.} SiH_3I + C_6H_6 \qquad 28$$

$$ClC_6H_4SiH_3 + HI \xrightarrow{\sim 20°C.} SiH_3I + C_6H_5Cl \qquad 29$$

$$(C_6H_5)_2SiH_2 + 2HI \xrightarrow{-40°C.} SiH_2I_2 + 2C_6H_6 \qquad 30$$

$$C_6H_5SiH_3 + HBr \xrightarrow{-78°C.} SiH_3Br + C_6H_6 \qquad 31$$

$$(C_6H_5)_2SiH_2 + HCl \xrightarrow{-78°C.} C_6H_5SiH_2Cl + C_6H_6 \qquad 32$$

In general, it appears that the more electronegative the halogen attached to the silicon, or the more halogens there are attached to the silicon, the slower is the cleavage reaction. Thus $C_6H_5SiCl_3$ is not cleaved by hydrogen halides. Also hydrogen chloride does not cleave

---

* R. D. Verma and L. C. Leitch [*Can. J. Chem.* **41**, 1652 (1963)] describe important improvements in the method of synthesizing SiH_3Br and SiH_3Cl from $C_6H_5SiH_3$ and the corresponding hydrogen halide.

the silicon–phenyl bond as readily as hydrogen bromide or hydrogen iodide. Aluminum chloride greatly catalyzes the reaction of hydrogen chloride with $C_6H_5SiH_3$, even at $-78\,°C$., but the SiH_3Cl formed cannot be isolated in the pure state since the aluminum chloride causes it to disproportionate.

Anhydrous hydrogen fluoride does not react at $-60\,°C$. with C_6H_5-SiH_3 probably because these substances form two immiscible liquid layers. However, quantitative conversion to benzene occurs at $0\,°C$. and volatile silicon fluorides are also formed. No pure fluoride has been isolated from the reaction. It appears possible that the SiH_3F which may be presumed to be first formed, disproportionates to other fluorides such as SiH_2F_2, etc., in the presence of the hydrogen fluoride. This is suggested by the fact that $SiHF_3$ although completely stable when pure, decomposes in the presence of hydrogen fluoride (85). In the presence of copper salts such as copper(II) chloride, fluorination of the silicon–hydrogen bonds is favored at $0\,°C$., thus $C_6H_5SiH_3$ gives good yields of phenyl fluorosilanes such as $C_6H_5SiF_3$ and the cleavage reaction is reduced.

There is evidence to suggest that mixed halosilanes such as SiH_2BrI may be formed in reactions of the type

$$C_6H_5SiH_2I + HBr \quad \rightarrow \quad SiH_2BrI + C_6H_6 \qquad \textbf{33}$$

However, no compound such as this has been isolated and characterized and it appears possible that halogen exchange,

$$SiH_2BrI + HBr \quad \rightarrow \quad SiH_2Br_2 + HI \qquad \textbf{34}$$

or disproportionation, may result in these compounds being very difficult to isolate and purify.

It should be noted that if hydrogen iodide is passed through boiling $C_6H_5SiH_3$ (b.p. $120\,°C$.) iodination of the silicon hydrogen bond takes place

$$C_6H_5SiH_3 + HI \quad \rightarrow \quad C_6H_5SiH_2I + H_2 \qquad \textbf{35}$$

in addition to the cleavage reaction given by eq. 28. The cleavage reaction is favored by lower temperatures and at $-40\,°C$. it is quantitative during one hour.

H. From Redistribution Reactions

It has been known for many years that silicon–halogen and silicon–hydrogen bonds will undergo redistribution reactions on heating at relatively low temperatures in the presence of aluminum halide catalyst. Silane and SiH_2Cl_2, for example, on heating at $100\,°C$. for seven days in the presence of aluminum chloride react to give an equilibrium mixture of chlorosilanes from which large yields of SiH_3Cl (35) can be obtained if appropriate proportions of the reactants are employed:

$$SiH_2Cl_2 + SiH_4 \rightarrow 2SiH_3Cl \qquad\qquad 36$$

Similarly SiH_4 and $SiCl_4$ give SiH_2Cl_2 on heating at 300 °C. with this catalyst (86). Under similar conditions any one chlorosilane will rearrange to give a mixture of all the possible disproportionation products.

Other catalysts besides aluminum halides are also effective in this type of reaction. For example, good yields of SiH_2Cl_2 may be obtained by refluxing $SiHCl_3$ (one of the few commercially available compounds containing silicon–hydrogen linkages) in the presence of dimethylcyanamide, for 5 hr. (87):

$$2SiHCl_3 \rightarrow SiH_2Cl_2 + SiCl_4 \qquad\qquad 37$$

At 200 °C., in the presence of the same catalyst, SiH_2Cl_2 gives 15% yields of SiH_3Cl (87). Dichlorosilane may also be obtained in 15% yield by heating $SiHCl_3$ to 300–400 °C. in the presence of acid catalysts of the metallic type such as Al_2Cl_6, Fe_2Cl_6, and BF_3 (88). It appears that this method could be appropriately modified to produce SiH_3I from $SiHI_3$, which can itself be conveniently prepared from $SiHCl_3$ (69).

Analogous types of reactions have also been employed to synthesize silicon tetrahalides in which two different halogen atoms are present in the molecule. This is illustrated by the good yields of $SiCl_3Br$, $SiCl_2Br_2$, and $SiClBr_3$ which are obtained when a mixture of $SiCl_4$ and $SiBr_4$ are passed through a tube at 600 °C. in the absence of catalyst (89). In a somewhat similar type of reaction it has been observed that at high temperatures the halogen atom of a gaseous hydrogen halide may replace a halogen in a silicon tetrahalide. Thus, $SiICl_3$ and hydrogen iodide react when heated in a bomb at 240–250 °C. according to the equation below (90):

$$SiICl_3 + HI \rightarrow SiI_2Cl_2 + HCl \qquad\qquad 38$$

III. DERIVATIVES OF SILANE

A. Fluorosilanes

SiH_3F (b.p. -88.1 °C.) (75). Fluorosilane (silyl fluoride) appears to be relatively stable thermally when pure, but in the presence of small amounts of unknown impurities it decomposes at a measurable rate at room temperature (16,63).

It may be prepared by the general method involving fluorination of SiH_3Cl with SbF_3 in the presence of a small amount of $SbCl_5$ catalyst.

This occurs readily in the vapor phase and after 24 hr. at room temperature a yield of 79% has been reported (16). However, it could not be obtained in a state of purity greater than 98% by this method. This was probably caused by slow disproportionation during the purification process since a sample slowly increased in vapor pressure when stored at room temperature.

The pure material may be conveniently obtained by the action of BF_3, at low temperatures, on amines (75) or ethers (63,65) containing the SiH_3 group, as previously described. Since $(SiH_3)_2O$ can be prepared more conveniently than an amine such as $(SiH_3)_3N$ [by, for example, the almost quantitative hydrolysis of SiH_3I (66)], the best practical method for synthesizing SiH_3F appears to be that involving the interaction of $(SiH_3)_2O$ with BF_3. The method reported by Onyszchuk (63) is described below. All manipulations were carried out in a vacuum system.

Boron trifluoride (0.0581 g.; 0.85 mmole) and $(SiH_3)_2O$ (0.1010 g.; 1.28 mmole) are condensed together and held at $-78°C.$ for 12 hr. Impure silyl fluoride is then distilled from the reaction vessel which is maintained at $-78°C.$ Final purification is effected by distilling it from a trap at $-132°C.$ (Yield: 0.0378 g.; 1.44 mmole). Since the amount of SiH_3F obtained is greater than that expected from the reaction given by eq. **16** it appears that some of the SiH_3OBF_2 first formed had undergone decomposition to liberate more SiH_3F according to eq. **17**. The white, solid SiH_3OBF_2 remaining in the reaction vessel decomposes gradually without melting when it is warmed to room temperature and distilled from trap to trap. The total yield of pure SiH_3F (0.110 g.; 2.18 mmole; mol. wt., found: 50.4, calc.: 50.1) is 85% based on the overall reaction:

$$3(SiH_3)_2O + 2BF_3 \rightarrow 6SiH_3F + B_2O_3 \qquad \textbf{39}$$

Analogous reactions to the above have proved successful in synthesizing CH_3SiH_2F and $(CH_3)_2SiHF$ from $(CH_3SiH_2)_2O$ and $[(CH_3)_2SiH]_2O$, respectively (61).

SiH_2F_2 (m.p. $-122.0°C.$; b.p. $-77.8°C.$). Difluorosilane appears to be a relatively stable substance and it has been shown to undergo no change in vapor pressure when stored over mercury for 14 days at room temperature. However, after 2 years it had completely disproportionated to SiH_4 and SiF_4 (16).

It may be obtained in yields up to 80% by fluorinating gaseous SiH_2Cl_2 with SbF_3 (containing 10% $SbCl_5$ catalyst) during 24 hr. at room temperature. Partially fluorinated products such as SiH_2FCl may also be formed in this reaction but they have not been isolated (16).

$SiHF_3$ (m.p. $-131.4°C.$; b.p. $-95.0°C.$). Trifluorosilane has been

reported to disproportionate (presumably to SiH_4 and SiF_4) at very low temperatures (54). Later work indicated that it was stable at low temperatures but disproportionated steadily at room temperature (16). However, it has recently been shown that when the compound is free of hydrogen fluoride it is stable for long periods even at elevated temperatures (85). In fact, complete disproportionation can only be made to occur by passing it over heated sodium fluoride (85):

$$4SiHF_3 + 6NaF \rightarrow SiH_4 + 3Na_2SiF_6 \qquad\qquad 40$$

Trifluorosilane may be prepared by the high temperature reaction of SiF_4 with hydrogen (85) or by the fluorination of $SiHCl_3$ by SbF_3 (54). Experimental details of the former method have not been published and the latter method gives poor yields since a large amount of the starting material is always recovered unchanged. The compounds $SiHFCl_2$ and $SiHF_2Cl$ are also obtained during the reaction of $SiHCl_3$ with SbF_3 (54). The most practical method for preparing relatively pure $SiHF_3$ involves the fluorination of $SiHCl_3$ in the vacuum system. Trichlorosilane vapor is passed backwards and forwards at room temperature for 3 hr. over SbF_3 containing about 5% $SbCl_5$ catalyst (16). Twenty-eight per cent of the $SiHCl_3$ was recovered unchanged in such an experiment, but 75% of the trichlorosilane consumed was converted to $SiHF_3$. With longer contact times, complete consumption of $SiHCl_3$ occurs and 91% yields of $SiHF_3$ can be obtained.

SiHFCl$_2$ (m.p. $-149.5\,°C.$; b.p. $-18.4\,°C.$). Fluorodichlorosilane may be prepared in relatively large quantities together with $SiHF_3$ and small amounts of $SiHF_2Cl$ during the fluorination of $SiHCl_3$ with SbF_3 (54). Antimony trifluoride is added slowly to $SiHCl_3$ containing 3–7% of $SbCl_5$ catalyst. A vigorous exothermic reaction occurs.

SiHF$_2$Cl (m.p. above $-144\,°C.$; b.p. $\sim -50\,°C.$). Difluorochlorosilane is obtained in small quantities during the preparation of $SiHF_3$ and $SiHFCl_2$ previously described (54).

B. Chlorosilanes

SiH$_3$Cl (m.p. $-118.1\,°C.$; b.p. $-30.4\,°C.$). Chlorosilane (silyl chloride) appears to be a reasonable stable compound (4,35). It may be prepared by heating a 10% excess of SiH_4 with hydrogen chloride in the presence of a small amount of aluminum chloride catalyst for 30 hr. at $100\,°C.$ Silyl chloride and SiH_2Cl_2 are formed in the molar ratio 4:1 (35). (See eq. 4.) However, since SiH_3I can now be readily prepared in large quantities it seems that a more simple method is to pass the vapor of SiH_3I over silver chloride:

$$SiH_3I + AgCl \rightarrow SiH_3Cl + AgI \qquad\qquad 41$$

This method has been found to give reasonably good yields of pure SiH_3Cl (53) when the reaction is carried out in a similar manner to that used in the preparation of SiH_3NCS described in Sec. III.F.3. It would seem possible that this conversion could be performed on the bench, using a solution of SiH_3I in an inert solvent to reduce the vigor of the reaction.

SiH_2Cl_2 (m.p. $-122\,°C$.; b.p. $8.3\,°C$.). Dichlorosilane appears to be completely stable where pure (4,35). It may be prepared by heating SiH_3Cl with hydrogen chloride in the presence of aluminum chloride catalyst or by heating SiH_4 and hydrogen chloride in the molar ratio $1:2$ in the presence of aluminum chloride for 7 days at $100\,°C$. In this way the chief product obtained is SiH_2Cl_2 (35). However, in view of the simple method now available for preparing SiH_2I_2 in relatively large quantities, it would appear to be preferable to synthesize it by passing the vapor of SiH_2I_2 over silver chloride as described in Sect. III.F.3 for the preparation of SiH_3NCS. Reasonably good yields of SiH_2Cl_2 have been obtained from SiH_2I_2 (50) by this method. It would seem possible that this conversion could be performed on the bench, using a solution of SiH_2I_2 in an inert solvent in order to moderate the reaction. If large quantities of SiH_2Cl_2 are required, it would probably be best prepared by the catalytic disproportionation of $SiHCl_3$ discussed in Sec. II.H (87,88). The $SiHCl_3$ itself may be purchased or prepared in large quantities from silicon and hydrogen chloride.

$SiHCl_3$ (m.p. $-126.5\,°C$.; b.p. $31.8\,°C$.). Trichlorosilane appears to be a completely stable compound (4,91). Although it is formed from SiH_4 and hydrogen chloride when they are heated in the presence of aluminum chloride catalyst, it is much more conveniently synthesized from silicon or ferrosilicon and hydrogen chloride at elevated temperatures (16,17,24,92). The method described by Whitmore et al. is given below (24).

The reaction is carried out in a horizontal, electrically heated, Pyrex tube (3-cm. diameter; 45-cm. length) the entrance of which is connected to a sulfuric acid drying tower and a calibrated flow meter. The exit is attached to a water-cooled condenser, which leads to a flask cooled in a salt–ice mixture. As a precaution against loss of material, the flask is connected to a trap immersed in a dry ice–acetone mixture which, in turn, leads to a sulfuric acid drying tower. The tube is filled with 375 g. granular ferrosilicon (ca. 90% Si), and the system is dried by passing nitrogen through it for 24 hr. at a furnace temperature of $300\,°C$. Dry hydrogen chloride is then passed in at $350\,°C$. until liquid first condenses in the receiver. The temperature is then lowered and held at 290–$310\,°C$. while the flow of hydrogen chloride is maintained at 0.6–0.85 mole/hour.

Dry nitrogen is passed through for 10 min. at approximately 10-hr. intervals. In 60 hr., 1370 g. of liquid products were obtained in a typical experiment. Fractionation through a column of approximately 15 theoretical plates gave 1045 g. SiHCl$_3$ (b.p. 31.5–32 °C./729 mm.). This represents 77% of the crude product, the remainder being chiefly SiCl$_4$ and small quantities of chloropolysilanes. In other similar experiments yields of 70–77% were obtained. It should be noted that a more rapid flow of hydrogen chloride at a reaction temperature of 360–370 °C. gave yields as low as 32%.

C. Bromosilanes

SiH_3Br (m.p. −94 °C.; b.p. 1.9 °C.). Bromosilane (silyl bromide) is spontaneously inflammable in air but appears to be thermally stable when pure (34).

It can be prepared in the usual way by heating silane with anhydrous hydrogen bromide in the presence of aluminum bromide catalyst (94). This method has been adapted to produce SiH$_3$Br in a continuous manner in any desired quantity by streaming the reactants through a furnace containing the catalyst (93). However, the labor involved in constructing the all-metal apparatus required is probably too great unless large quantities of the compound are needed. Small quantities of SiH$_3$Br may be obtained by brominating SiH$_4$ with solid bromine at −80 °C. Some SiH$_2$Br$_2$ is also formed in this reaction (34).

The best method for preparing SiH$_3$Br would appear to be through the reaction of excess hydrogen bromide with C$_6$H$_5$SiH$_3$ at −78 °C. (See eq. 31.) This proceeds quantitatively during a 3-hr. period (80,82). Care should be exercised because of the spontaneously inflammable nature of SiH$_3$Br.

SiH_2Br_2 (m.p. −70.1 °C.; b.p. 66 °C.). Dibromosilane appears to be a stable material but it is sometimes spontaneously inflammable in air (34). It may be obtained in the pure state by the reaction of SiH$_4$ with hydrogen bromide in the presence of aluminum bromide catalyst on gentle heating (93,94). It is also formed by the reaction of solid bromine with SiH$_4$ at −80 °C. (34).

If a solution of SiH$_2$Br$_2$ in benzene is satisfactory then it is most easily prepared by reacting C$_6$H$_5$SiH$_2$Br with excess hydrogen bromide at −78 °C. for 6 days (82):

$$C_6H_5SiH_2Br + HBr \rightarrow SiH_2Br_2 + C_6H_6 \qquad \textbf{42}$$

The C$_6$H$_5$SiH$_2$Br required can be prepared from (C$_6$H$_5$)$_2$SiH$_2$ and HBr. Since benzene has a boiling point (80.1 °C.) somewhat close to that of

SiH_2Br_2, some difficulty may be experienced in separating the substances unless a good fractionating column is employed.

A possible alternative method for preparing pure SiH_2Br_2 might be the reaction of SiH_2I_2 with silver bromide. The Conversion Series discussed previously suggests that such a reaction should take place.

SiHBr₃ (m.p. $-73.5°C.$; b.p. $111.8°C.$). Tribromosilane is an apparently stable material but it is usually spontaneously inflammable when poured through air. Although it can be prepared by brominating SiH_4 at low temperatures with elemental bromine (34), it is more practical to prepare it by heating silicon at $360-400°C.$ in a stream of hydrogen bromide in an analogous method to that used for preparing $SiHCl_3$ (25,26).

D. Iodosilanes

SiH₃I (m.p. $-57.0°C.$; b.p. $45.4°C.$). Iodosilane (silyl iodide) is a reasonably stable material and it can be stored at room temperature in sealed tubes without much decomposition occurring (36). Silyl iodide can be prepared by heating SiH_4 with hydrogen iodide in the presence of aluminum iodide catalyst at $80-100°C.$ for 12–18 hr. (36,53). More completely iodinated compounds such as SiH_2I_2, $SiHI_3$, and SiI_4 are also formed depending on the relative amounts of reactants employed.

It can now be prepared very much more conveniently by cleaving either $C_6H_5SiH_3$ or $ClC_6H_4SiH_3$ with hydrogen iodide (see eqs. **28** and **29**). Quantitative yields of SiH_3I (based on $C_6H_5SiH_3$ used) have been reported in the former case (37) and 60% yields (based on $C_6H_5SiCl_3$ used) have been reported in the latter (84).

The method described by Fritz and Kummer (37) for preparing SiH_3I from $C_6H_5SiH_3$ and hydrogen iodide is given below.

Phenylsilane, $C_6H_5SiH_3$, is prepared by *slowly* adding commercial $C_6H_5SiCl_3$ (150 g.), with constant stirring, to $LiAlH_4$ (23 g.) in 700 ml. diethyl ether. The mixture is then refluxed for 3 hr. The solution, which contains dissolved lithium and aluminum salts, is filtered through a sintered glass funnel with exclusion of air. The ether and $C_6H_5SiH_3$ are then distilled from the dissolved salts at reduced pressure, care being taken not to allow the temperature of the flask to rise above $30°C.$ If distillation is carried out at atmospheric pressure, the resulting higher temperature required may cause the $C_6H_5SiH_3$ to disproportionate in the presence of the dissolved aluminum chloride (95) to give SiH_4 which is spontaneously inflammable in air and can cause serious explosions:

$$4C_6H_5SiH_3 \rightarrow 3SiH_4 + (C_6H_5)_4Si \qquad \textbf{43}$$

The salt-free solution of $C_6H_5SiH_3$ can be distilled safely at atmospheric pressure to give pure $C_6H_5SiH_3$ (b.p. 120 °C.) (yield: 61 g., 79%).

A molar excess of hydrogen iodide is then condensed on to any desired quantity of $C_6H_5SiH_3$. After standing at -40 °C. for 1 hr., mixing is facilitated by warming the materials and then recondensing them. This is repeated two or three times. The excess hydrogen iodide is allowed to evaporate and the products are separated by distillation.

A variation of the above method has been reported by Aylett and Ellis (84). Chlorophenylsilane, $ClC_6H_4SiH_3$, is used instead of C_6H_5-SiH_3 since the chlorobenzene formed in the reaction (see eq. **29**) has a higher boiling point than benzene and is, therefore, easier to separate from the SiH_3I. The synthesis, which is carried out in the vacuum system, is as follows:

Phenyltrichlorosilane, $C_6H_5SiCl_3$ (25 g.), is purified by fractional condensation at -46 °C. and is then converted to $ClC_6H_4SiCl_3$ (96,97). [Chlorination readily occurs at 60–70 °C. in 40 min. in the presence of 0.5% iron powder as catalyst (97).] Twenty-five ml. anhydrous diethyl ether is added to the yellowish liquid product obtained and the solution is then added slowly to a mixture of $LiAlH_4$ (16 g.) and ether (30 ml.) contained in a nitrogen-filled three-necked flask fitted with a reflux condenser and guard tube. Reaction is very vigorous. After standing overnight, all volatile materials are removed from the reaction flask by pumping for 6 hr. The flask is occasionally warmed during this period. The volatile material obtained is distilled several times through a -64 °C. trap in which the $ClC_6H_4SiH_3$ condenses. The chlorophenyl-silane is placed in a thick-walled Pyrex tube (approximately 100 ml.) and anhydrous hydrogen iodide (25 g.) is distilled in. After sealing, the tube is allowed to warm slowly to room temperature and after 12 hr. the volatile products present are fractionated to yield pure SiH_3I (yield: 11 g., 60%; mol. wt., found: 158, calcd.: 157.9). (*Note:* Several distillations through a trap held at -64 °C. should be sufficient to remove all chlorobenzene from the more volatile SiH_3I which passes through a trap held at this temperature.)

SiH_2I_2 (m.p. -1.0 °C.; b.p. 149.5 °C.). Diiodosilane appears to be a stable material (36). It may be prepared from SiH_4 and hydrogen iodide as described for SiH_3I (36), but it is more conveniently obtained in high yields by reacting $(C_6H_5)_2SiH_2$ with hydrogen iodide (82,84) (see eq. **30**) in a method analogous to that given previously for the preparation of SiH_3I from $C_6H_5SiH_3$.

$SiHI_3$ (m.p. 8 °C.; b.p. 220 °C.). Triiodosilane is a stable material which is formed in small quantities during the reaction of SiH_4 with hydrogen iodide described earlier (36). It may be obtained by passing

hydrogen iodide over heated silicon (27) in a preparation analogous to that used for $SiHCl_3$ and $SiHBr_3$. The best method of synthesis uses $SiHCl_3$ as a starting material (69). The $SiHCl_3$ is converted to $(SiHN)_x$ by ammonia or to $SiH(NHC_6H_5)_3$ by aniline. Both these compounds are readily decomposed by hydrogen iodide to give good yields of $SiHI_3$ (72) (see eqs. **22** and **23**). The method described by Ruff (69) is given below.

Dry aniline (125 g.) is dissolved in dry benzene (300 ml.) in a 1-l. suction flask, the side arm of which is attached to a drying tube. Trichlorosilane (25 g.) dissolved in dry benzene (150 ml.) is then added slowly through a dropping funnel inserted through a stopper in the neck of the flask. The contents of the flask should be vigorously agitated by swirling or by a magnetic stirrer throughout the addition which should take approximately 1 hr. The temperature gradually rises to 50–60°C. (see eq. **22**). The precipitated $C_6H_5NH_2 \cdot HCl$ is filtered off with exclusion of moisture in an atmosphere of dry air, nitrogen, or carbon dioxide and is washed three times with 50 ml. portions of benzene.

The benzene solution of $SiH(NHC_6H_5)_3$ is then placed in another suction flask equipped with a drying tube on the side arm and a stream of dry hydrogen iodide is slowly bubbled through the solution by way of a tube inserted through a stopper in the neck of the flask. The reaction given by eq. **23** takes place. Passage of hydrogen iodide is discontinued when copious fumes caused by hydrogen iodide are observed coming from the drying tube. The precipitated $C_6H_5NH_2 \cdot HI$ is filtered with exclusion of moisture. After washing with three 50-ml. portions of benzene, the benzene is removed by distillation at approximately 50 mm. pressure (40–50°C.). If the distilled benzene is more than a light pink in color, it may be inferred that distillation is too rapid and that some of the product is also distilling over. The residue, which is deep red in color because of the presence of free iodine, is shaken with a little mercury to remove the iodine. If this is not removed, the yield of $SiHI_3$ will be reduced because of the reaction

$$SiHI_3 + I_2 \rightarrow SiI_4 + HI \qquad \textbf{44}$$

which tends to occur during the subsequent distillation. The crude $SiHI_3$ is placed in a smaller flask and is distilled at approximately 20 mm. pressure (110°C.), (Yield: 45–50 g.; 60–65%).

E. Tetrahalosilanes

SiF_4 (m.p. −90.2°C./1318 mm.; sublm. pt. −95.7°C.). Silicon tetrafluoride is a thermally stable gas which is formed when hydrofluoric acid or hydrogen fluoride comes in contact with substances containing SiO_2. Depending on the exact method of preparation, the SiF_4 is usually contaminated to varying extents with hydrogen fluoride or H_2SiF_6(98). It may be prepared in a state of high purity by heating $BaSiF_6$ at temperatures between 300 and 500 °C.(98). The $BaSiF_6$ itself can be conveniently prepared from barium chloride and aqueous fluorosilicic acid.

$SiCl_4$ (m.p. −68 °C.; b.p. 57.0 °C.) (99). Silicon tetrachloride is a thermally stable material which may be prepared in a variety of ways. When relatively large quantities are required the most practical method is to pass chlorine or dry hydrogen chloride over heated silicon or silicon alloy (or metal silicide) (7,24,100–103). Best yields are obtained at higher temperatures. At lower temperatures the proportion of higher chlorides such as Si_2Cl_6, Si_3Cl_8, etc., formed increases. It appears that the initial compounds produced are probably the most complex and contain chains of Si–Si atoms such as are present in the element itself. In the presence of excess chlorine, and aided by higher temperatures, these chains are progressively broken down to give finally $SiCl_4$.

$SiBr_4$ (m.p. 5.4 °C.; b.p. 154 °C.). Silicon tetrabromide is thermally stable and may be most conveniently prepared by passing bromine vapor in a stream of nitrogen over heated granular silicon. Under appropriate flow and heating conditions, essentially pure $SiBr_4$ is obtained as the only product (25,26,100,104–106).

SiI_4 (m.p. 120.5 °C.; b.p. 287.5 °C.). Silicon tetraiodide is thermally stable but turns dark in color when in a sealed tube in direct sunlight (100,107). It is interesting to note that SiI_4 is actually more stable than its carbon analog, CI_4, which decomposes to C_2I_4 both in the presence of heat and sunlight (108). It may be conveniently prepared as a colorless crystalline material by passing iodine vapor in a current of carbon dioxide over heated silicon (106,107).

SiF_3Cl (m.p. −142.0 °C.; b.p. −70.0 °C.). Trifluorochlorosilane may be prepared by fluorinating $SiCl_4$ with SbF_3. The $SiCl_4$ is mixed with a deficit of SbF_3 and $SbCl_5$ catalyst is then added slowly to the mixture. Good yields of SiF_3Cl in addition to SiF_2Cl_2 and $SiFCl_3$ are obtained from the exothermic reaction which takes place (56,109). The compound is also formed by mixing Si_2F_6 with chlorine. Upon heating the reaction vessel strongly with a flame at one point, a mild explosion occurs and both SiF_3Cl and SiF_2Cl_2 are produced (32). Trifluorochlorosilane has also been prepared in small yields, together with SiF_2Cl_2 and $SiFCl_3$, by passing a mixture of SiF_4 and $SiCl_4$ vapor through a tube at 700 °C. (110).

SiF_2Cl_2 (m.p. −139.7 °C.; b.p. −32.2 °C.). Difluorodichlorosilane may be obtained as described above for SiF_3Cl (32,56,109,110).

$SiFCl_3$ (m.p. −120.8 °C.; b.p. 12.2 °C.). Fluorotrichlorosilane may be prepared as described above for SiF_3Cl (32,56,109,110).

SiF₃Br (m.p. 70.5 °C.; b.p. −41.7 °C.). Trifluorobromosilane has been obtained by slowly adding SbF_3 (no catalyst) to heated $SiBr_4$. Good yields of SiF_3Br in addition to SiF_2Br_2 and $SiFBr_3$ are obtained (111). It may also be prepared, together with SiF_2Br_2 and $SiFBr_3$, by passing a mixture of SiF_4 and $SiBr_4$ vapor through a tube at 700 °C. (110).

SiF₂Br₂ (m.p. −66.9 °C.; b.p. 13.7 °C.). Difluorodibromosilane shows a tendency to decompose on standing. It may be conveniently obtained in good yields as described for SiF_3Br (110,111). The compound can also be synthesized, together with SiF_3Br, by mixing Si_2F_6 with excess bromine and then heating the container at one point with a small flame. A minor explosion occurs to give the above products (111).

SiFBr₃ (m.p. −82.5 °C.; b.p. 83.8 °C.). Fluorotribromosilane can be obtained readily as described for SiF_3Br (110,111).

SiFCl₂Br (m.p. −112.3 °C.; b.p. 35.4 °C.). Fluorodichlorobromosilane may be prepared by slowly adding SbF_3 to $SiCl_2Br_2$ (112). It may also be obtained by heating $SiFBr_3$ with chlorine in a sealed tube for 2 hr. with a small flame or by heating $SiFBr_3$ with $SbCl_3$ (112).

SiFClBr₂ (m.p. −99.3 °C.; b.p. 59.5 °C.). Fluorochlorodibromosilane is formed when SbF_3 is gradually added, with rapid stirring, to $SiClBr_3$ or by heating $SiFBr_3$ with $SbCl_3$ (112).

SiF₃I (m.p. ∼ −92 °C.; b.p. −24 °C.). Trifluoroiodosilane is not very stable to distillation at atmospheric pressure. It is prepared together with SiF_2I_2 and $SiFI_3$ by passing SiF_4 mixed with SiI_4 vapor through a tube at 700 °C. Reaction occurs without the use of a catalyst and random distribution of halogen atoms occurs (110).

SiF₂I₂ (m.p. ∼ −83 °C.; b.p. 84.5 °C.). Difluorodiiodosilane is prepared as described for SiF_3I (110).

SiFI₃ (m.p. ∼ −41 °C.; b.p. 188 °C.). Fluorotriiodosilane may be prepared as described for SiF_3I (110).

SiCl₃Br (b.p. 80 °C.). Trichlorobromosilane may be synthesized together with $SiCl_2Br_2$ and $SiClBr_3$ by: (a) passing a mixture of chlorine and bromine over heated silicon at 700 °C.; (b) refluxing $SiBr_4$ in an atmosphere of chlorine in a sealed tube for 48 hr.; (c) heating a mixture of $SiBr_4$ and $SbCl_3$; or (d) passing a mixture of the vapors of bromine and Si_2Cl_6 (or Si_3Cl_8) through a tube at 500 °C. (112).

SiCl₂Br₂ (b.p. 105 °C.). Dichlorodibromosilane can be obtained by the methods given above for $SiCl_3Br$ (112).

SiClBr₃ (b.p. 128 °C.). Chlorotribromosilane may also be prepared as described for $SiCl_3Br$ (112).

SiCl₃I (b.p. 113.5 °C.). Trichloroiodosilane is a heavy colorless liquid which decomposes slowly at room temperature with liberation of iodine. Decomposition is accelerated by light (90). It is conveniently prepared by heating SiHCl$_3$ and iodine in a bomb at 240–250°C. for 15 hr. (90, 113):

$$SiHCl_3 + I_2 \rightarrow SiCl_3I + HI \qquad \textbf{45}$$

A large amount of SiCl$_2$I$_2$ is also formed according to eq. **38**. Trichloroiodosilane may also be prepared together with SiCl$_2$I$_2$ and SiClI$_3$ by passing a mixture of hydrogen iodide and SiCl$_4$ vapor through a red hot glass tube (113). It may be obtained in small quantities by treating SiCl$_4$ with fused potassium iodide (89).

SiCl$_2$I$_2$ (b.p. 172°C.). Dichlorodiiodosilane is similar to SiCl$_3$I in its chemical properties. It may be formed by the methods given for SiCl$_3$I (90,113)or by passing SiCl$_3$I vapor through a heated tube (89).

SiClI$_3$ (m.p. 2°C.; b.p. 234–237°C.). Chlorotriiodosilane may be prepared from SiCl$_4$ and hydrogen iodide (113) as described for SiCl$_3$I, or by heating SiCl$_3$I (89).

SiBr$_3$I (m.p. 14°C.; b.p. 192°C.). Tribromoiodosilane is a colorless solid which may be prepared together with small amounts of SiBr$_2$I$_2$ and SiBrI$_3$ by heating SiHBr$_3$ and iodine in a sealed tube at 200–250°C. It may also be obtained, mixed with larger quantities of the other two bromoiodosilanes, by passing IBr vapor over silicon at red heat (114,115).

SiBr$_2$I$_2$ (m.p. 38°C.; b.p. 230.5°C.). Dibromodiiodosilane can be obtained by the methods given for the preparation of SiBr$_3$I (114,115).

SiBrI$_3$ (m.p. 53°C.; b.p. ∼255°C.). Bromotriiodosilane is prepared by the methods given above for SiBr$_3$I (114,115).

F. Halogenoid Derivatives

1. Cyanides

SiH$_3$CN (m.p. 32.4°C.; b.p. 49.6°C.) (49). Silyl cyanide has been proven to have the normal cyanide structure (116–118), but there is some question as to the structure of substituted silyl cyanides such as (CH$_3$)$_3$SiCN (2). It has been suggested that this latter compound may actually be an equilibrium mixture of the normal and the isocyanides (73,119).

Silyl cyanide has an unexpectedly high melting point and is a colorless crystalline solid at room temperature and can be readily sublimed *in vacuo*. If completely pure, samples of SiH$_3$CN may be stored in sealed tubes at room temperature for 6 months with little or no decomposition occurring. However, in the presence of traces of impurities such as condensed mercury vapor, decomposition proceeds rapidly in a few days at room temperature.

Silyl cyanide may be prepared in good yields by pumping the vapor of SiH$_3$I over silver cyanide at room temperature (49)(see eq. **10**). Analogous compounds such as CH_3SiH_2CN may be synthesized in a similar manner (120).

Cl_3SiCN (m.p. $-46.2\,°C$.; b.p. $73.2\,°C$.). Trichlorocyanosilane has been obtained by the interaction of mercuric cyanide with liquid or gaseous Si_2Cl_6 at $100\,°C$. (121).

2. Isocyanates

SiH_3NCO (m.p. $-88.6\,°C$.; b.p. $18.1\,°C$.). Silyl isocyanate, like other silicon cyanates, is believed to have the iso structure (48). It differs from CH_3NCO but is similar to SiH_3NCS in that it is a linear (symmetric top) molecule. When pure it is stable for at least 2 weeks at room temperature, but in the presence of small amounts of impurities it rapidly decomposes.

It is synthesized by bubbling nitrogen through SiH_3I at $-46\,°C$. The gaseous nitrogen–SiH_3I mixture is then pumped through silver cyanate:

$$SiH_3I + AgNCO \rightarrow SiH_3NCO + AgI \qquad \textbf{46}$$

It is of particular interest to note that attempts to prepare SiH_3NCO by pumping the vapor of SiH_3I *alone* through silver cyanate were unsuccessful (49). Complex decomposition occurred with the formation of Si-$(NCO)_4$. It has now been shown that silver cyanate catalytically decomposes SiH_3NCO at room temperature, but that dilution of the SiH_3I vapor with nitrogen during the reaction retards this decomposition sufficiently to permit the isolation of SiH_3NCO although the yields (up to 25%) are not large (48).

SiF_3NCO (b.p. $-6.0\,°C$.). Trifluoroisocyanatosilane is thermally unstable at room temperature and tends to disproportionate at its boiling point. It is prepared by slowly adding antimony trifluoride to boiling $Si(NCO)_4$ (122). No catalyst is required. It is obtained in good yields together with relatively large amounts of $SiF_2(NCO)_2$ and $SiF(NCO)_3$. It may also be prepared together with $SiF_2(NCO)_2$ and $SiF(NCO)_3$ by passing a mixture of SiF_4 and $Si(NCO)_4$ vapor through a tube at $700\,°C$. (110).

$SiF_2(NCO)_2$ (m.p. $-75.0\,°C$.; b.p. $68.6\,°C$.). Difluorodiisocyanatosilane is prepared together with SiF_3NCO as described above (110,122).

$SiF(NCO)_3$ (m.p. $-29.2\,°C$.; b.p. $134.3\,°C$.). Fluorotriisocyanatosilane is obtained from $Si(NCO)_4$ as described for SiF_3NCO (110,122).

$SiCl_3NCO$ (m.p. $-69\,°C$.; b.p. $86.8\,°C$.). Trichloroisocyanatosilane may be prepared together with $SiCl_2(NCO)_2$ and $Si(NCO)_4$ by very slowly adding silver cyanate to a large excess of $SiCl_4$ dissolved in a mixture of organic solvents. The reaction is exothermic. It may also be synthesized together with $SiCl_2(NCO)_2$ and $SiCl(NCO)_3$ by a redistribution reaction. Silicon tetrachloride and Si(NCO)$_4$ are heated together in a Carius tube for 70 hr. at 135 °C. Good yields of all products are obtained. It has been calculated that a period of 1 week at 135 °C. would be required for complete random distribution to be obtained. The products are formed very much more rapidly if the vapors of SiCl$_4$ and Si(NCO)$_4$ are passed through a hot tube at 600 °C. (123).

SiCl$_2$(NCO)$_2$ (m.p. −80 °C.; b.p. 117.8 °C.). Dichlorodiisocyanato-silane is obtained by the methods described for SiCl$_3$NCO (123). It may also be formed in a general type of reaction in which the most volatile component distils from the system. Thus, when Si(NCO)$_4$ and C$_6$H$_5$COCl (or a similar high-boiling chloride) are heated, SiCl$_2$(NCO)$_2$ and SiCl(NCO)$_3$ may be distilled from the mixture (124).

SiCl(NCO)$_3$ (m.p. −35 °C.; b.p. 152 °C.). Chlorotriisocyanatosilane is obtained by the methods described for SiCl$_3$NCO (123) and SiCl$_2$-(NCO)$_2$ (124).

Si(NCO)$_4$ (m.p. 26.0 °C.; b.p. 185.6 °C.). Raman and infrared studies show that the compound definitely has the isocyanate structure (125). Silicon tetraisocyanate is readily prepared by adding SiCl$_4$ to a suspension of silver cyanate in benzene. Reaction is completed by heating the mixture for half an hour on a steam bath (126). Contrary to earlier reports no silicon tetracyanate Si(OCN)$_4$, is formed in this reaction (127). It may also be prepared by treating potassium cyanate dissolved in a mixture of fused lithium and potassium chlorides with SiCl$_4$ (128).

3. Isothiocyanates

SiH$_3$NCS (m.p. −51.8 °C.; b.p. 84.0 °C.). Silyl isothiocyanate, like other silicon thiocyanates, appears to have the iso structure (129,130). It is similar to SiH$_3$NCO, but differs from CH$_3$NCS in having a linear (symmetric top) structure (129–132). Silyl isothiocyanate decomposes during 2–3 weeks at room temperature.

It may be prepared by pumping the vapor of SiH$_3$I over silver thiocyanate at room temperature:

$$SiH_3I + AgNCS \quad \rightarrow \quad SiH_3NCS + AgI \qquad \textbf{47}$$

The method described by MacDiarmid for its synthesis is given below (49).

In a typical preparation, a glass reaction tube containing silver thiocyanate (20 g.) mixed with glass wool is attached to the vacuum system by way of a trap which can later be immersed in liquid nitrogen. A tube containing SiH$_3$I (1.1560 g.) is connected to the other end of the reaction tube by means of a cone and socket joint and a stopcock. The apparatus is pumped for 2–3 hr. to remove traces of adsorbed water on the silver thiocyanate and glass wool. The trap is then immersed in liquid nitrogen and the SiH₃I is slowly pumped through the reaction tube and the products are condensed in the liquid nitrogen trap. It is necessary to pump constantly, otherwise the hydrogen liberated during the reaction (due to decomposition) will slow down the rate of flow of SiH_3I through the reaction tube enormously. A long contact time with the silver salt also generally decreases the yield of product. In order to reduce the amount of decomposition occurring during the reaction, the rate of flow of SiH_3I through the apparatus is controlled by adjusting the stopcock so that the reaction tube becomes barely warm to the touch. A period of 2 hr. was required to pass the SiH_3I through the reaction tube. If unreacted SiH_3I is found during the purification procedure below, it may be again treated with silver thiocyanate but a fresh charge of silver salt must be used. (Reaction appears to occur only on the surface of the silver thiocyanate particles, and the surface area available for reaction depends greatly on the degree to which the silver salt is evenly dispersed on the glass wool.)

The crude SiH_3NCS obtained is distilled several times through a trap held at −64 °C. in which it condenses. It is then separated from less volatile impurities by passing it through a trap held at −48 °C. (Yield: 0.4340 g.; 66%; mol. wt., found: 89.5; calcd.: 89.2.)

$SiCl_3NCS$ (m.p. −75 °C.; b.p. 129.5 °C.). Trichloroisothiocyanatosilane is relatively stable thermally but less so than $SiCl_3NCO$. It rearranges to $Si(NCS)_4$ and $SiCl_4$ to an extent of approximately 35% during 4 months at room temperature, whereas $SiCl_3NCO$ undergoes not more than 2–3% rearrangement under comparable conditions. It is prepared by adding an approximately equimolar quantity of silver thiocyanate to $SiCl_4$ in six separate portions, with intermittent half-hour periods of reflux on a steam bath (133,142).

$Si(NCS)_4$ (m.p. 143.8 °C.; b.p. 313.0 °C.) (133). Silicon tetraisothiocyanate is a thermally stable, colorless, crystalline solid which may be obtained by adding $SiCl_4$ to a suspension of lead thiocyanate in benzene. Reaction is brought to completion by heating on a water bath (143). It may also be prepared by reacting a mixture of fused sodium and potassium thiocyanates with $SiCl_4$ (128). Raman and infrared studies show that it definitely has the isothiocyanate structure (134).

4. Isoselenocyanates

SiH_3NCSe (m.p. −15.1 °C.; b.p. ∼111 °C.). Silyl isoselenocyanate polymerizes slowly at room temperature, even in the dark. Heat, ultraviolet light, AgNCSe, or traces of BF_3 accelerate decomposition. It is prepared by passing SiH_3I vapor, diluted with nitrogen, over AgNCSe at room temperature (48a).

5. Azides

SiH_3N_3 (m.p. $-81.8\,°C.$; b.p. $\sim28\,°C.$). Silyl azide is sufficiently stable to isolate and characterize (71), but it decomposes slowly in the vacuum system with liberation of SiH_4. It is interesting to note that organosilicon azides appear to be surprisingly stable thermally (58,135). Thus, $(CH_3)_3SiN_3$ requires temperatures as high as $500\,°C.$ to induce thermal decomposition (58).

Silyl azide is prepared by treating $(SiH_3)_3N$ or $C_2H_5N(SiH_3)_2$ with a solution of HN_3 in di-n-butyl ether (71):

$$(SiH_3)_3N + 4HN_3 \rightarrow 3SiH_3N_3 + NH_4N_3 \qquad \textbf{48}$$

$Si(N_3)_4$. Silicon tetraazide is a highly explosive white crystalline solid which is sensitive to moisture. It is soluble in benzene and ether. It has been obtained in a slightly impure state by refluxing a solution of $SiCl_4$ in benzene with sodium azide in the presence of catalytic quantities of $LiAlH_4$ (57).

IV. DERIVATIVES OF DISILANE

A. Fluorodisilanes

Si_2H_5F (m.p. $-100.4\,°C.$; b.p. $-10.0\,°C.$). Disilanyl fluoride is probably spontaneously inflammable in air but it shows no sign of decomposition when held in the vapor phase for 25 hr. at $0\,°C.$ It may be prepared in 78% yield by the reaction of $(Si_2H_5)_3N$ with BF_3 at $-78\,°C.$ (76):

$$(Si_2H_5)_3N + BF_3 \rightarrow Si_2H_5F + (Si_2H_5)_2NBF_2 \qquad \textbf{49}$$

Si_2F_6 (m.p. $-18.7\,°C./780$ mm.; sublm. p. $-19.1\,°C.$). Hexafluorodisilane is a stable substance which sublimes on warming from low temperatures. It may be easily prepared in good yields by warming Si_2Cl_6 to $50\text{-}60\,°C.$ with anhydrous zinc fluoride. A vigorous reaction occurs which, once started, continues without further application of heat for an hour or more. The Si_2F_6 evolved can be conveniently collected in a trap immersed in an acetone–dry ice bath (136).

B. Chlorodisilanes

Si_2H_5Cl (m.p. $-111.6\,°C.$; b.p. $40.1\,°C.$). Disilanyl chloride is spontaneously inflammable in air and shows only very slight thermal decomposition at $18\,°C.$ However, at $-24\,°C.$ it undergoes rapid decomposition in the liquid phase in the presence of traces of hydrogen chloride or aluminum chloride (51).

It may be prepared by the reaction of excess Si_2H_6 with hydrogen chloride in the presence of aluminum chloride catalyst at room temperature in a reaction analogous to that given by eq. 5 (39). However, because of the decomposition already mentioned, the yields obtained are very small and this reaction is not recommended as a preparative method. It is more conveniently synthesized in high yields by pumping the vapor of Si_2H_5I over silver chloride (51) in a method analogous to that used for the preparation of SiH_3NCS described in Sec. III.F. 3.

$Si_2H_xCl_{6-x}$. More completely chlorinated disilanes are formed in abundance in the reaction of Si_2H_6 with hydrogen chloride described above; however, none of these have been identified (39). Chlorodisilanes are also formed by the little understood reaction of Si_2H_6 with carbon tetrachloride or chloroform in the presence of traces of air (43). The compound, Si_2Cl_5H, appears to be formed when a mixture of hydrogen and $SiCl_4$ vapor is subjected to a glow discharge (137).

Si_2Cl_6 (m.p. 2.5 °C.; b.p. 147 °C.) (103,138). Hexachlorodisilane is relatively stable thermally. It is obtained together with $SiCl_4$ when chlorine is passed over heated silicon alloys or silicides (103). It can also be prepared from $SiCl_4$ by various electrical discharge methods in the presence of a halogen acceptor such as hydrogen or a metal (103,138). Alternatively, it can be formed by the reaction of $SiCl_4$ with silicon at 1000 °C. (103). It is best prepared, together with chlorides of the higher silanes, by passing a *slow* stream of chlorine over a calcium silicon alloy ("calcium silicide"), at a *low* temperature (150 °C.). Higher flow rates and higher temperatures greatly reduce the yield of product and increase the amount of $SiCl_4$ formed (103). Under the above optimum conditions, the product obtained has the following approximate composition: 65% $SiCl_4$, 30% Si_2Cl_6, 4% Si_3Cl_8, 1% Si_4Cl_{10}, Si_5Cl_{12}, and Si_6Cl_{14} (103).

C. Bromodisilanes

Si_2H_5Br (m.p. −97.2 °C.; b.p. 69.5 °C.). Disilanyl bromide is spontaneously inflammable in air and shows only slight decomposition on heating to 50 °C. The pure compound, or a mixture with hydrogen bromide, undergoes very little decomposition on standing at 0 °C. for 2–3 days; however, rapid decomposition occurs at 0 °C. in the presence of aluminum bromide (139).

Disilanyl bromide may be prepared by pumping the vapor of Si_2H_5I over silver bromide (139), but it is best prepared in relatively large quantities by the reaction of Si_2H_6 with hydrogen bromide in the presence of aluminum bromide catalyst either at room temperature or at $-78\,°C$. (39). The reaction which occurs is analogous to that given by eq. **5**. Disilanyl bromide is probably the most convenient starting material to use in the synthesis of compounds containing the Si_2H_5 group. It is preferred to Si_2H_5I since it can be distilled more readily in the vacuum system and larger quantities of the bromide can be synthesized in a given time.

The method of preparation described by Abedini et al. (39) is given below.

In a typical preparation, disilane (7.84 g., 126 mmole) (140) and hydrogen bromide (3.4 g., 42 mmole) are condensed in a side-arm tube connected to a 5-l. reaction bulb which is attached to a vacuum system. The reaction bulb contains anhydrous aluminum bromide ($\sim$0.5 g.) sublimed on its inner walls. The reactants are allowed to vaporize into the reaction bulb and after 5 min. at room temperature all volatile material is pumped from the reaction bulb through four traps immersed in liquid nitrogen in order to remove the hydrogen which is formed. Unreacted Si_2H_6 and hydrogen bromide (if any) are removed from the products by distillation through a trap held at $-96\,°C.$, in which the products only condense. The products are then transferred to a low-temperature fractionating column (described below), and upon distillation a small amount of Si_2H_6 (column temperature $\sim -96\,°C.$) is first obtained and then pure Si_2H_5Br (column temperature $\sim -64\,°C.$) (1.08 g., 7.68 mmole; 18% yield based on hydrogen bromide; mol. wt., found: 141.4, calcd.: 141.1). At somewhat warmer column temperatures, slightly impure Si_2H_5Br (0.72 g., 5.1 mmole; 12% yield; mol. wt., found: 145.0) is obtained. This impure material can be used as such for most syntheses requiring Si_2H_5Br (Total yield: 30%). Very little, if any, unreacted hydrogen bromide is recovered from this reaction and no cleavage of the silicon–silicon bond has ever been observed.

The low-temperature fractionating column used in the above synthesis is not essential, but it gives a more rapid separation and a larger amount of pure Si_2H_5Br than conventional trap-to-trap distillations. It is also useful in other vacuum system separations. It consists of a length of Dewar-walled tubing (inside diameter: 72 mm.; outside diameter: 100 mm.; length: 250 mm.). A large cork is placed in each end of the tube. A 600-mm. length of glass tubing (inside diameter: 15 mm.) is inserted through the corks. A small glass bulb (diameter: 30 mm.) is attached to one end of the tube so that the distance between the bottom of the bulb and the lower end of the Dewar-walled tube is 150 mm. A ground glass joint is fitted to the other end of the tube so that the assembly may be attached to the vacuum system. Two lengths of 8-mm.

tubing and a pentane thermometer are inserted through the top cork. One of the 8-mm. lengths of tubing is attached to a Dewar (25 or 50 l.) containing liquid nitrogen. A length of heating tape or a small electric heater is dropped into the liquid nitrogen. On passing a current through the heater, the nitrogen is evaporated and cold nitrogen gas enters the still and escapes through the second length of 8-mm. tubing. The greater the current passed through the heater, the colder the temperature in the still becomes. With a 50-l. Dewar, temperatures in the column will remain within $\pm 1°$ at $\sim -130°C.$ for an hour or more. The temperature observed on the pentane thermometer is only approximate and varies with the positioning of the thermometer. A small magnetic stirring bar can be placed in the bulb at the end of the column if it is considered desirable to stir the contents of the bulb. This is often the case if large amounts of material are being distilled.

To operate the fractionating column, material is distilled into the bulb by immersing it in liquid nitrogen. The column temperature is then made as low as possible. The liquid nitrogen Dewar is removed from the bulb and the column temperature is increased until a fraction starts to distil out. The temperature is then held at this value until no more material distils. The column temperature is then increased until the next fraction begins to distil out and is then held constant at this value. This procedure is repeated for each new fraction which distils.

$Si_2H_xBr_{6-x}$. Large quantities of polybromodisilanes are obtained during the reaction of Si_2H_6 with hydrogen bromide previously described (39), but none of these have been investigated.

Si_2Br_6 (m.p. 95°C.; b.p. 265°C.). Hexabromodisilane is relatively stable thermally. It is a colorless, crystalline material which is soluble in a variety of organic solvents (105). Although it may be prepared by the action of bromine on silicon, the use of a calcium-silicon alloy instead of silicon permits a lower reaction temperature and materially increases the yield of the desired product (105). In this respect the synthesis is analogous to that of Si_2Cl_6; however, it differs from that of Si_2Cl_6 in that oxygen can be used to carry the bromine vapor over the heated silicon (at 180–200°C.) and practically no silicon oxybromides are found in the product (12,105).

D. Iododisilanes

Si_2H_5I (m.p. −86.1°C.; b.p. 102.8°C.). Disilanyl iodide is spontaneously inflammable in air and it undergoes fairly rapid decomposition at room temperature in a few hours (38). It seems likely that this decomposition is catalyzed by traces of aluminum iodide impurity which is difficult to remove. It may be prepared by the reaction of Si_2H_6 with hydrogen iodide in the presence of aluminum iodide catalyst at room temperature (38) as given by eq. **5**.

$Si_2H_xI_{6-x}$. Large quantities of polyiododisilanes are obtained during the reaction of Si_2H_6 with hydrogen iodide previously described (38), but none of these have been investigated.

Si_2I_6 [m.p. (decomp.) 250°C.]. Hexaiododisilane begins to undergo decomposition at approximately 250°C. When heated slowly at 350–400°C. in a stream of nitrogen it partly decomposes to give SiI_4 and solid $(SiI)_x$ (107). The thermal stability of the hexahalodisilanes decreases on progressing from the chloride to the iodide (141). Hexaiododisilane may be prepared by heating SiI_4 with finely divided metallic silver at 280°C. for 6 hr. (100,107):

$$2SiI_4 + 2Ag \quad \rightarrow \quad Si_2I_6 + 2AgI \qquad \qquad \textbf{50}$$

E. Halogenoid Derivatives

No stable inorganic halogenoid derivative of disilane or of the higher silanes has been isolated. Thus, attempts to prepare $Si_2(NCO)_6$ were unsuccessful (13) and Si_2H_5CN decomposes rapidly at low temperatures (51). This, perhaps, is not altogether surprising, since even $(CH_3)_3$-$SiSi(CH_3)_2CN$ decomposes on refluxing (144):

$$x(CH_3)_3SiSi(CH_3)_2CN \quad \rightarrow \quad (x-1)(CH_3)_3SiCN + (CH_3)_3Si[Si(CH_3)_2]_xCN \quad \textbf{51}$$

V. DERIVATIVES OF THE HIGHER SILANES

Relatively little is known about halogen derivatives of the higher silanes. Several molecular species such as Si_4Cl_{10}, which are volatile or soluble in organic solvents, have been isolated, but until relatively recently they have not been studied in any detail. A fairly large number of silicon "subhalides" such as $(SiBr)_n$ or $(SiBr_2)_n$ have been isolated. These are frequently amorphous, insoluble, infusable substances and may be regarded as halogen derivatives of the corresponding silanes $(SiH)_n$ and $(SiH_2)_n$. Compounds of the type $(SiX_2)_n$ are, in effect, halogenated higher silanes of general formula $X_3Si(SiX_2)_nSiX_3$ where n is very large, or they may be cyclic species. When there are less than two halogen atoms per silicon atom then crosslinking between silicon atoms may be expected to occur with resulting decrease in volatility and solubility of the compound in organic solvents (2).

A. Fluorosilanes

No definite derivatives of the higher silanes have been isolated (7).

B. Chlorosilanes

$Si_3H_4Cl_4$ has been prepared from Si_3H_8 and chloroform (43) as previously described (See eq. **9**).

Si_3Cl_8 (m.p. $-67\,°C.$, b.p. $216\,°C.$) is obtained during the preparation of Si_2Cl_6 as previously described (7,103).

Si_4Cl_{10} (b.p. $150\,°C./15$ mm.), Si_5Cl_{12} (b.p. $190\,°C./15$ mm.), and Si_6Cl_{14} (sub. *in vacuo*, $200\,°C.$) obtained during the preparation of Si_2Cl_6 described earlier (7,103) probably consist of a mixture of isomers and they have not been well characterized (103). When heated, the vapors of the higher chlorides inflame in air.

Si_5Cl_{12} [m.p. (decomp.) $341\,°C.$] has been prepared by heating Si_6Cl_{14} with trimethylamine at $70\,°C.$ It is believed that it may have a neopentyl type of structure (145,146).

Si_6Cl_{14} (m.p. $320\,°C.$) has been prepared in a most interesting reaction involving the trimethylamine induced disproportionation of Si_2Cl_6 (145,146) which takes place readily at $0\,°C.$:

$$5Si_2Cl_6 \quad \rightarrow \quad Si_6Cl_{14} + 4SiCl_4 \qquad\qquad \mathbf{52}$$

It is believed that this compound might have a neo-hexyl type of structure (145,146).

$Si_{10}Cl_{22}$ is a very viscous oil which can be distilled at $215\text{–}220\,°C.$ *in vacuo*. Its identity has been confirmed by molecular weight measurements. The compound is inflammable and it decomposes when heated in a vacuum to give silicon, $SiCl_4$, and Si_2Cl_6. It is prepared by passing a mixture of hydrogen and $SiCl_4$ vapor over a quartz rod heated to $1000\text{–}1100\,°C.$ and then condensing the gaseous products in liquid air (147).

$Si_{25}Cl_{52}$ is a solid which ignites in air on scratching. It is soluble in ether. The compound is prepared in a manner similar to $Si_{10}Cl_{22}$ except that nitrogen is substituted for the hydrogen (148).

$Si_{10}Cl_{20}H_2$ is a colorless, extremely viscous oil which decomposes on heating to give $(SiCl)_x$. It is prepared by strongly heating a mixture of hydrogen and $SiCl_4$ vapor in a "hot–cold" tube (149,150).

$(SiCl_2)_x$ is a white solid which can be obtained by subjecting a mixture of hydrogen and $SiCl_4$ vapor to a glow discharge (137) or by passing chlorine over heated silicon under controlled conditions (151).

$[SiCl_{(0.5-2.6)}]_x$. Compounds of the composition range indicated may be prepared by the action of a glow discharge on $SiHCl_3$ or a mixture of hydrogen and $SiCl_4$ vapor (152). $[SiCl_{2.6}]_x$ may be obtained as a viscous, distillable oil by passing $SiCl_4$ vapor over silicon at $1000\,°C.$ (153).

$(SiCl)_x$ is a yellow solid obtained by decomposing $Si_{10}Cl_{20}H_2$ in an inert atmosphere at 300 °C. (150,154).   It is also formed by prolonged heating of Si_2Cl_6 (141).

C. Bromosilanes

Si_3Br_8 (m.p. 133 °C.) and Si_4Br_{10} [m.p. (decomp.) 185 °C.] are prepared by the action of an electric discharge on the vapor of $SiHBr_3$ (155).

$(SiBr_2)_x$ is a brown, rosin-like solid which is soluble in nonpolar organic solvents.   It ignites at temperatures above 100 °C. and on heating to 300 °C. it decomposes to $(SiBr)_x$ and Si_2Br_6.   It is formed by the reduction of $SiBr_4$ with silicon at 1200 °C. (156).

$(SiBr)_x$ is a brownish-yellow solid which is insoluble in benzene.   It is prepared by refluxing Si_2Br_6 at 300 °C. for 6–8 hr. (141).   It may also be obtained by reacting magnesium with $SiBr_4$ in ether (156).

D. Iodosilanes

$(SiI)_x$ is an orange-red amorphous solid which is insoluble in all organic solvents and does not melt.   It is prepared by slowly heating Si_2I_6 at 350–400 °C. in a stream of nitrogen (107).

REFERENCES

1. Stone, F. G. A., *Hydrogen Compounds of the Group IV Elements*, Prentice-Hall, Englewood Cliffs, New Jersey, 1962.
1a. Ebsworth, E. A. V., *Volatile Silicon Compounds*, Pergamon Press, New York, 1963.
2. MacDiarmid, A. G., "Silanes and Their Derivatives," in *Advances in Inorganic Chemistry and Radiochemistry*, Vol. III, H. J. Eméleus and A. G. Sharpe, eds., Academic Press, New York, 1961, p. 207.
3. MacDiarmid, A. G., *Quart. Revs. (London)* **10**, 208 (1956).
4. Stock, A., *Hydrides of Boron and Silicon*, Cornell Univ. Press, Ithaca, New York, 1933.
5. Borer, K., and C. S. G. Phillips, *Proc. Chem. Soc.* **1959**, 189.
6. Eaborn, C., *Organosilicon Compounds*, Butterworths, London, 1960.
7. *Gmelins Handbuch der Anorganischen Chemie*, Vol. 15, Verlag Chemie, Weinheim, 1959, Part B.
8. *Gmelins Handbuch der Anorganischen Chemie*, Vol. 15, Verlag Chemie, Weinheim, 1958, Part C.
9. Rochow, E. G., *An Introduction to the Chemistry of the Silicones*, 2nd ed., Wiley, New York, 1951.
10. Post, H. W., *Organic Silicon Compounds*, Reinhold, New York, 1949.
11. Rochow, E. G., D. T. Hurd, and R. N. Lewis, *The Chemistry of Organometallic Compounds*, Wiley, New York, 1957.
12. Schumb, W. C., and C. H. Klein, *J. Am. Chem. Soc.* **59**, 261 (1937).
13. Forbes, G. S., and H. H. Anderson, *J. Am. Chem. Soc.* **69**, 3048 (1947).
14. Pflugmacher, A., and H. Dahmen, *Z. Anorg. Allgem. Chem.* **290**, 184 (1957).

15. Panckhurst, D. J., C. J. Wilkins, and P. W. Craighead, *J. Chem. Soc.* **1955**, 3395.
16. Emeléus, H. J., and A. G. Maddock, *J. Chem. Soc.* **1944**, 293.
17. Booth, H. S., and W. D. Stillwell, *J. Am. Chem. Soc.* **56**, 1529 (1934).
18. Jolly, W. L., *Synthetic Inorganic Chemistry*, Prentice-Hall, Englewood Cliffs, New Jersey, 1960, Chap. 10.
19. Sanderson, R. T., *Vacuum Manipulation of Volatile Compounds*, Wiley, New York, 1948.
20. Dodd, R. E., and P. L. Robinson, *Experimental Inorganic Chemistry*, Elsevier, New York, 1954, Chap. 2.
21. Stock, A., *Hydrides of Boron and Silicon*, Cornell Univ. Press, Ithaca, New York, 1933, Chap. 30.
22. Dushman, S., and J. M. Lafferty, *Scientific Foundations of Vacuum Technique*, Wiley, New York, 1949.
23. Farkas, A., and H. W. Melville, *Experimental Methods in Gas Reactions*, Macmillan, London, 1939.
24. Whitmore, F. C., E. W. Pietrusza and L. H. Sommer, *J. Am. Chem. Soc.* **69**, 2108 (1947).
25. Schumb, W. C., and R. C. Young, *J. Am. Chem. Soc.* **52**, 1464 (1930).
26. Schumb, W. C., in *Inorganic Syntheses*, Vol. 1, H. S. Booth, ed., McGraw-Hill, New York, 1939, p. 38.
27. Buff, H., and F. Wöhler, *Annalen*, **104**, 94 (1857).
28. Taylor, A. G., and B. V. de G. Walden, *J. Am. Chem. Soc.* **66**, 842 (1944).
29. Combes, C., *Compt. Rend.* **122**, 531 (1896).
30. Eaborn, C., *Organosilicon Compounds*, Butterworths, London, 1960, pp. 36–45.
31. British Thomson-Houston Co., British Patent 626,519 (1949); through *Chem. Abstr.* **44**, 2547 (1950).
32. Schumb, W. C., and E. L. Gamble, *J. Am. Chem. Soc.* **54**, 3943 (1932).
33. Eaborn, C., *Organosilicon Compounds*, Butterworths, London, 1960, pp. 209–212.
34. Stock, A., and C. Somieski, *Chem. Ber.* **50**, 1739 (1917).
35. Stock, A., and C. Somieski, *Chem. Ber.* **52B**, 695 (1919).
36. Emeléus, H. J., A. G. Maddock, and C. Reid, *J. Chem. Soc.* **1941**, 353.
37. Fritz, G., and D. Kummer, *Z. Anorg. Allgem. Chem.* **304**, 322 (1960).
38. Ward, L. G. L., and A. G. MacDiarmid, *J. Am. Chem. Soc.* **82**, 2151 (1960).
39. Abedini, M., C. H. Van Dyke, and A. G. MacDiarmid, *J. Inorg. Nucl. Chem.* **25**, 307 (1963).
40. Anderson, H. H., *J. Am. Chem. Soc.* **80**, 5083 (1958).
41. Anderson, H. H., and A. Hendifar, *J. Am. Chem. Soc.* **81**, 1027 (1959).
42. Anderson, H. H., *J. Am. Chem. Soc.* **81**, 4785 (1959).
43. Stock, A., and P. Stiebeler, *Chem. Ber.* **56B**, 1087 (1923).
44. Emeléus, H. J., and S. R. Robinson, *J. Chem. Soc.* **1947**, 1592.
45. Anderson, H. H., and H. Fischer, *J. Org. Chem.* **19**, 1296 (1954).
46. Eaborn, C., *J. Chem. Soc.*, **1950**, 3077.
47. Eaborn, C., *Organosilicon Compounds*, Butterworths, London, 1960, p. 174.
48. Ebsworth, E. A. V., and M. J. Mays, *J. Chem. Soc.* **1962**, 4844.
48a. Ebsworth, E. A. V., and M. J. Mays, *J. Chem. Soc.* **1961**, 4879; **1963**, 3893.
49. MacDiarmid, A. G., *J. Inorg. Nucl. Chem.* **2**, 88 (1956).
50. Abedini, M., Ph.D. Dissertation, University of Pennsylvania, 1963.
51. Craig, A. D., J. V. Urenovitch, and A. G. MacDiarmid, *J. Chem. Soc.* **1962**, 548.

52. Emeléus, H. J., A. G. MacDiarmid, and A. G. Maddock, *J. Inorg. Nucl. Chem.* **1**, 194 (1955).
53. Abedini, M., and A. G. MacDiarmid, unpublished results, 1963.
54. Booth, H. S., and W. D. Stillwell, *J. Am. Chem. Soc.* **56**, 1531 (1934).
55. Ruff, O., and C. Albert, *Chem. Ber.* **38**, 53, 222 (1905).
56. Booth, H. S., and C. F. Swinehart, *J. Am. Chem. Soc.* **57**, 1333 (1935).
57. Wiberg, E., and H. Michaud, *Z. Naturforsch.* **9b**, 500 (1954).
58. Connolly, J. W., and G. Urry, *Inorg. Chem.* **1**, 718 (1962).
59. Hurd, D. T., *J. Am. Chem. Soc.* **67**, 1545 (1945).
60. Cottrell, T., *The Strength of Chemical Bonds*, 2nd ed., Butterworths, London, 1958, pp. 270–283.
61. Emeléus, H. J., and M. Onyszchuk, *J. Chem. Soc.* **1958**, 604.
62. Eaborn, C., *Organosilicon Compounds*, Butterworths, London, 1960, pp. 171–173.
63. Onyszchuk, M., *Can. J. Chem.* **39**, 808 (1961).
64. Crane, E. J., *Chem. Eng. News* **24**, 1233 (1946).
65. Sternbach, B., and A. G. MacDiarmid, *J. Am. Chem. Soc.* **83**, 3384 (1961).
66. Kriner, W. A., A. G. MacDiarmid, and E. C. Evers, *J. Am. Chem. Soc.* **80**, 1546 (1958).
37. Orlov, N. F., *Doklady Akad. Nauk SSSR* **114**, 1033 (1957); through *Chem. Abstr.* **52**, 2742 (1958).
68. Stock, A., and C. Somieski, *Chem. Ber.* **54**, 740 (1921).
69. Ruff, O., *Chem. Ber.* **41**, 3738 (1908).
70. Anderson, H. H., *J. Am. Chem. Soc.* **73**, 5802 (1951).
71. Ebsworth, E. A. V., D. R. Jenkins, M. J. Mays, and T. M. Sugden, *Proc. Chem. Soc.* **1963**, 21.
72. Birkofer, L., A. Ritter, and P. Richter, *Angew. Chem. Intern. Ed. in Engl.* **1**, 267 (1962).
73. Bither, T. A., W. H. Knoth, R. V. Lindsey, Jr., and W. H. Sharkey, *J. Am. Chem. Soc.* **80**, 4151 (1958).
74. Burg, A. B., and E. S. Kuljian, *J. Am. Chem. Soc.* **72**, 3103 (1950).
75. Sujishi, S., and S. Witz, *J. Am. Chem. Soc.* **79**, 2447 (1957).
76. Abedini, M., and A. G. MacDiarmid, *Inorg. Chem.* **2**, 608 (1963).
77. Ebsworth, E. A. V., and H. J. Emeléus, *J. Chem. Soc.* **1958**, 2150.
78. Sujishi, S., *Chemistry of Group IV Hydrides*, Final Report, Office of Ordnance Research, Contract No. DA-11-022-ORD-1264, Aug., 1957.
79. Becke-Goehring, M., and H. Krill, *Chem. Ber.* **94**, 1059 (1961).
79a. Fritz, G., *Z. Anorg. Allgem. Chem.* **280**, 332 (1955).
80. Fritz, G., and D. Kummer, *Z. Anorg. Allgem. Chem.* **308**, 105 (1961).
81. Fritz, G., and D. Kummer, *Z. Anorg. Allgem. Chem.* **310**, 327 (1961).
82. Fritz, G., and D. Kummer, *Chem. Ber.* **94**, 1143 (1961).
83. Fritz, G., and D. Kummer, *Z. Anorg. Allgem. Chem.* **306**, 191 (1960).
84. Aylett, B. J., and I. A. Ellis, *J. Chem. Soc.* **1960**, 3415.
85. Wolfe, J. K., and N. C. Cook, Abstr. Papers, 128th National Meeting of the American Chemical Society, Minneapolis, Minn., 1955, p. 48M.
86. Clasen, H., *Angew. Chem.* **70**, 179 (1958).
87. Bailey, D. L., and G. H. Wagner, U. S. Patent 2,732,280–2 (1956); through *Chem. Abstr.* **50**, 12097 (1956).
88. Erickson, C. E., and G. H. Wagner, U. S. Patent 2,627,451 (1953); through *Chem. Abstr.* **48**, 1420 (1954); U. S. Patent 2,735,861 (1956); through *Chem. Abstr.* **50**, 13986 (1956).

89. Forbes, G. S., and H. H. Anderson, *J. Am. Chem. Soc.* **66,** 931 (1944).
90. West, R., and E. G. Rochow, in *Inorganic Syntheses,* Vol. 4, J. Bailar, ed., McGraw-Hill, New York, 1953, p. 41.
91. Stock, A., and F. Zeidler, *Chem. Ber.* **56B,** 986 (1923).
92. Dudani, P. G., and H. G. Plust, *Nature* **194,** 85 (1962).
93. Opitz, H. E., J. S. Peake, and W. H. Nebergall, *J. Am. Chem. Soc.* **78,** 292 (1956).
94. Stock, A., and C. Somieski, *Chem. Ber.* **51,** 989 (1918).
95. Speier, J. L., and R. E. Zimmerman, *J. Am. Chem. Soc.* **77,** 6395 (1955).
96. Yakubovich, A. Y., and G. V. Motsarev, *Doklady Akad. Nauk SSSR* **91,** 277 (1953).
97. Yakubovich, A. Y., and G. V. Motsarev, *Zh. Obshch. Khim.* **26,** 568 (1956).
98. Hoffman, C. J., and H. S. Gutowsky, in *Inorganic Syntheses,* Vol. 4, J. Bailar, ed., McGraw-Hill, New York, 1953, p. 145.
99. *Gmelins Handbuch der Anorganischen Chemie,* Vol. 15, Verlag Chemie, Weinheim, 1959, Part B, pp. 672–673.
100. Schumb, W. C., *Chem. Rev.* **31,** 587 (1942).
101. Jackson, K. E., *Chem. Rev.* **25,** 67 (1939).
102. Baxter, G. P., P. F. Weatherill, and E. O. Holmes, Jr., *J. Am. Chem. Soc.* **42,** 1194 (1920).
103. Schumb, W. C., and E. L. Gamble, in *Inorganic Syntheses,* Vol. 1, H. S. Booth, ed., McGraw-Hill, 1939, p. 42.
104. Kennard, S. M. S., and P. A. McCusker, *J. Am. Chem. Soc.* **70,** 1039 (1948).
105. Schumb, W. C., in *Inorganic Syntheses,* Vol. 2, W. C. Fernelius, ed., McGraw-Hill, 1946, p. 98.
106. Gatterman, L., *Chem. Ber.* **22,** 186 (1889).
107. Schwarz, R., and A. Pflugmacher, *Chem. Ber.* **75B,** 1062 (1942).
108. McArthur, R. E., and J. H. Simons, in *Inorganic Syntheses,* Vol. 3, L. F. Audrieth. ed., McGraw-Hill, New York, 1950, p. 37.
109. Booth, H. S., and C. F. Swinehart, *J. Am. Chem. Soc.* **54,** 4750 (1932).
110. Anderson, H. H., *J. Am. Chem. Soc.* **72,** 2091 (1950).
111. Schumb, W. C., and H. H. Anderson, *J. Am. Chem. Soc.* **58,** 994 (1936).
112. Schumb, W. C., and H. H. Anderson, *J. Am. Chem. Soc.* **59,** 651 (1937).
113. Besson, A., *Compt. Rend.* **112,** 611, 1314 (1891).
114. Friedel, C., *Chem. Ber.* **2,** 57 (1869).
115. Besson, A., *Compt. Rend.* **112,** 788, 1447 (1891).
116. Linton, H. R., and E. R. Nixon, *Spectrochim. Acta* **10,** 299 (1958).
117. Muller, N., and R. C. Bracken, *J. Chem. Phys.* **32,** 1577 (1960).
118. Sheridan, J., and A. C. Turner, *Proc. Chem. Soc.* **1960,** 21.
119. Seyferth, D., and N. Kahlen, *J. Am. Chem. Soc.* **82,** 1080 (1960).
120. Emeléus, H. J., M. Onyszchuk, and W. Kuchen, *Z. Anorg. Allgem. Chem.* **283,** 74 (1956).
121. Kaczmarczyk, A., and G. Urry, *J. Am. Chem. Soc.* **81,** 4112 (1959).
122. Forbes, G. S., and H. H. Anderson, *J. Am. Chem. Soc.* **69,** 1241 (1947).
123. Anderson, H. H., *J. Am. Chem. Soc.* **66,** 934 (1944).
124. Anderson, H. H., *J. Am. Chem. Soc.* **75,** 1576 (1953).
125. Miller, F. A., and G. L. Carlson, *Spectrochim. Acta* **17,** 977 (1961).
126. Forbes, G. S., and H. H. Anderson, *J. Am. Chem. Soc.* **62,** 761 (1940).
127. Anderson, H. H., private communication, 1963.
128. Sundermeyer, W., *Z. Anorg. Allgem. Chem.* **313,** 290 (1961).
129. MacDiarmid, A. G., and A. G. Maddock, *J. Inorg. Nucl. Chem.* **1,** 411 (1955).

130. Jenkins, D. R., R. Kewley, and T. M. Sugden, *Proc. Chem. Soc.* **1960,** 220.
131. Jenkins, D. R., R. Kewley, and T. M. Sugden, *Trans. Faraday Soc.* **58,** 1284 (1962).
132. Ebsworth, E. A. V., R. Mould, R. Taylor, G. R. Wilkinson, and L. A. Woodward, *Trans. Faraday Soc.* **58,** 1069 (1962).
133. Anderson, H. H., *J. Am. Chem. Soc.* **69,** 3049 (1947).
134. Carlson, G. L., *Spectrochim. Acta* **18,** 1529 (1962).
135. West, R., and J. S. Thayer, *J. Am. Chem. Soc.* **84,** 1763 (1962).
136. Schumb, W. C., and E. L. Gamble, *J. Am. Chem. Soc.* **54,** 583 (1932).
137. Schwarz, R., and G. Pietsch, *Z. Anorg. Allgem. Chem.* **232,** 249 (1937).
138. Stock, A., A. Brandt, and H. Fischer, *Chem. Ber.* **58,** 643 (1925).
139. Ward, L. G. L., and A. G. MacDiarmid, *J. Inorg. Nucl. Chem.* **20,** 345 (1961).
140. Bethke, G. W., and M. K. Wilson, *J. Chem. Phys.* **26,** 1107 (1957).
141. Pflugmacher, A., and I. Rohrman, *Z. Anorg. Allgem. Chem.* **290,** 101 (1957).
142. Anderson, H. H., *J. Am. Chem. Soc.* **67,** 223 (1945).
143. Reynolds, J. E., *J. Chem. Soc.* **89,** 397 (1906).
144. Urenovitch, J. V., and A. G. MacDiarmid, *J. Am. Chem. Soc.* **85,** 3372 (1963).
145. a. Urry, G., and J. W. Nuss, Abstr. of Papers, 144th National Meeting of the American Chemical Society, Los Angeles, Calif., 1963, p. 16K; b. private communication, 1963.
146. Kaczmarczyk, A., M. Millard, and G. Urry, *J. Inorg. Nucl. Chem.* **17,** 186 (1961).
147. Schwarz, R., and H. Meckbach, *Z. Anorg. Allgem. Chem.* **232,** 241 (1937).
148. Schwarz, R., and C. Danders, *Chem. Ber.* **80,** 444 (1947).
149. Schwarz, R., and R. Thiel, *Z. Anorg. Allgem. Chem.* **235,** 247 (1938).
150. Schwarz, R., *Angew. Chem.* **51,** 328 (1938).
151. Antipin, P. F., and V. V. Sergeev, *Zh. Prik. Khim.* **27,** 784 (1954); through *Chem. Abstr.* **49,** 765 (1955).
152. Hertwig, K. A., and E. Wiberg, *Z. Naturforsch.* **6b,** 336 (1951).
153. Rochow, E. G., and R. Didtschenko, *J. Am. Chem. Soc.* **74,** 5545 (1952).
154. Schwarz, R., and U. Gregor, *Z. Anorg. Allgem. Chem.* **241,** 395 (1939).
155. Besson, A., and L. Fournier, *Compt. Rend.* **151,** 1055 (1910).
156. Schmeisser, M., and M. Schwarzmann, *Z. Naturforsch.* **11b,** 278 (1956).

CHAPTER 8

Saline Hydrides

CHARLES E. MESSER

Tufts University, Medford, Massachusetts

CONTENTS

I. INTRODUCTION

A. General

The majority of known metal hydrides are synthesized directly from metal plus hydrogen, with variations in procedure according to the thermodynamics and kinetics of the reaction. In the case of the saline hydrides—principally those of the alkali and alkaline earth metals—the reaction is, in general, straightforward, and only one hydride phase is formed in the final room temperature product. In the case of the metallic hydrides—those of the lanthanides, the actinides, the titanium and vanadium families, palladium, etc.—the procedure is often nearly the same. However, the composition and structure of the phases formed is apt to vary much more with conditions of preparation, so that detailed knowledge of the individual metal–hydrogen phase diagram is usually necessary for reliable synthesis. Thus this chapter is primarily concerned with the saline hydrides, which include the alkali metals, the alkaline earth metals except beryllium, and the lanthanides europium and ytterbium.

A brief description of the properties of the individual hydrides is given by Hurd (1). A thorough review and discussion of hydrides is given by Gibb (2).

B. Dissociation Pressure

The saline hydrides, on heating to higher temperatures, decompose reversibly into the metal and hydrogen. The metal phase is able to dissolve some hydrogen, and the hydride phase becomes hydrogen deficient. This mutual solubility, which varies from metal to metal, increases with increasing temperature, as does the hydrogen dissociation pressure.

This behavior is shown graphically in the series of pressure–composition isotherms of Figure 1 for a hypothetical metal–hydrogen system. On isothermal addition of hydrogen to metal, the gas dissolves in the metal until saturation is reached at the composition n' atoms H per atom metal and at the pressure p_1, at the temperature T_1. Further addition of hydrogen will then cause hydride formation. This will continue at the constant "plateau" pressure p_1 until the composition reaches n'' and the metal phase disappears. At this point, the hydride phase is saturated with hydrogen vacancies. Further addition of hydrogen continues with rapidly increasing pressure up to a limiting hydrogen absorption of nearly the stoichiometric s atoms H per atom metal, depending on the pressure used.

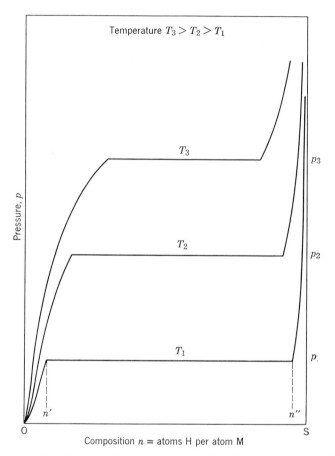

Fig. 1. Schematic diagram of pressure–composition isotherms for metal–metal hydride system M–MH$_s$.

As the temperature increases from T_1 to T_2 to T_3, the saturation limits n' and n'' approach each other, and the limiting composition of the hydride phase at a given hydrogen pressure—say 1 atm.—departs more and more from stoichiometry. Hence, synthesis of the purest possible hydride must be effected, or at least completed, at the lowest temperature permitted by the kinetics of the hydriding process.

II. APPARATUS AND TECHNIQUE

A. General

An apparatus suitable for the preparation of most metal hydrides in small quantities is shown in Figures 2, 3, and 4. It contains the usual

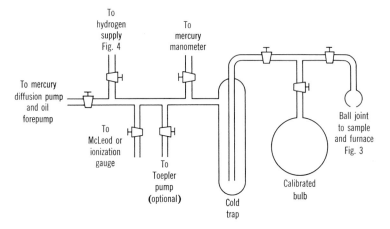

Fig. 2. Schematic diagram of vacuum and hydrogen measurement system.

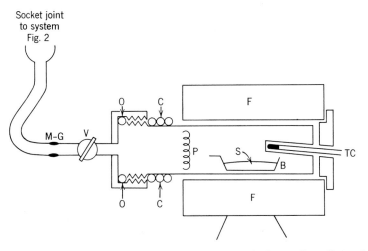

Fig. 3. Sample container and furnace arrangement.   B, boat; C, cooling coil; F, furnace; M–G, metal–glass seal; O, O-ring gasket; P, stainless steel wool plug; S, sample; TC, thermocouple; V, valve.

equipment of a glass high vacuum system, a hydrogen purification and metering system, and a sample container for the synthesis, with certain modifications and options.

The metal is weighed out into a sample boat in an argon-filled drybox. The boat is placed in the sample bomb, which is closed, removed from the drybox, attached to the vacuum system, and evacuated.   Purified hydrogen is added, and the amount measured by pressure–volume–tem-

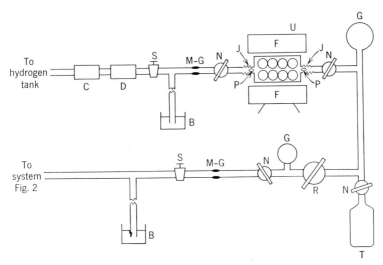

Fig. 4. Hydrogen introduction and purification system. B, mercury bubbler; C, catalytic purifier; D, drying tube; F, furnace; G, pressure gauge; J, threaded joint; M–G, metal-to-glass seal; N, needle valve; P, stainless steel wool plug; R, reducing valve; S, glass stopcock; T, high-pressure storage tank; U, uranium or titanium metal getter.

perature measurements. The furnace is then heated to cause reaction, the time–temperature pattern depending on the particular hydride. The amount of hydrogen unreacted is measured at some definite furnace temperature, so that the amount of hydrogen reacted can be calculated. It may be desirable to shut off the sample, add more measured hydrogen to the system, and reexpose the sample to hydrogen, to ensure a high final equilibrium pressure for maximum hydrogen uptake.

B. Sample Container

Normally, the outer sample container, Figure 3, is a horizontal tube fitted with a vacuum-tight joint and a cooling coil, which may be connected to the vacuum line by a ball-and-socket joint, with a valve or stopcock between the joints. The tube is provided with one or more thermocouple wells for temperature measurement and control. For small scale synthesis, the outer container may be 1 in. in diameter, and the furnace a standard 12 in. × $1^1/_4$ in. combustion furnace.

The outer tube may be of quartz for most hydride syntheses. For lithium hydride, stainless steel (type 347 or 316) is essential because of the rapid attack of glass or silica by lithium metal. The other metal vapors may attack silica (the author has observed this for barium), but do so only slowly.

The quartz outer container is opened for loading through a tapered ground glass joint. The stainless steel bomb is opened by means of a threaded joint with O-ring gasket. For a larger stainless steel bomb, the O-ring gasket may be replaced with a flanged joint with a flat ring gasket. A tested vacuum-tight needle valve is essential for the connecting tube of the stainless steel bomb.

A vertical furnace with crucible-type sample container could also be used. Single crystals of lithium hydride have been grown in this type of apparatus by the Stockbarger technique (3).

C. Hydrogen Purification

Figure 4 shows an apparatus for providing high purity hydrogen for hydride synthesis. Tank hydrogen is fed through a catalytic purifier, drying tube, and mercury safety "bubbler," to a steel tube filled with titanium or uranium metal and capable of standing high pressure. This tube, which is enclosed by a furnace, is heated to cause hydrogen uptake at 1 atm. pressure and a fairly low temperature to form titanium or uranium hydride. The hydrogen is then released at a higher temperature to a high pressure storage tank, from which it may be fed to the synthesis apparatus through a reducing valve and a second bubbler. The titanium or uranium acts as a getter for oxygen and nitrogen in the hydrogen, reacting with these but releasing only hydrogen on heating.

For preparations where the highest purity is not demanded, the getter and high pressure storage system may be omitted and the hydrogen may be fed directly to the second bubbler which is separated from the first bubbler and the vacuum line by means of stopcocks.

D. Materials and Materials Handling

All of these metals and their hydrides react readily with the constituents of air, so that "anaerobic" techniques are essential. An argon-filled drybox is essential for most operations, although in earlier years small scale work was done by other means (4,5). A large drybox of this sort is described by Gibb (6). For most operations, a smaller commercially available box might be used, providing that complete protection from air is assured by evacuation of box or entrance lock. The size of the entrance lock is critical in the design of all equipment to contain the hydride. The need to evacuate box or lock is critical in planned technique —everything going into the box must be capable of being evacuated, or must be in containers capable of withstanding 1 atm. of internal pressure. No wood, paper, or other moisture-containing material may be used.

The metal to be hydrided is removed from its original container in the drybox if possible. Otherwise, it is trimmed of its excess oxide–nitride coating externally under oil, rinsed with anhydrous ether, and transferred to the drybox lock wet with ether. It is then fully trimmed of its coating in the box. Most of the metals are capable of such trimming by a knife or simple tool which will go into the box.

In many cases, the metal is sufficiently pure for use in this way. Where it is not, vacuum distillation in a stainless steel apparatus is necessary. Douglas et al. (7) describe a suitable vacuum still.

The sample container for the metal during the synthesis is usually in the form of a combustion boat. For an outer container of 1-in. diameter, it may be a standard combustion boat size of $3^1/_2$-in. long by $^5/_8$-in. wide by $^3/_8$-in. deep. It may be of Armco iron, stainless steel type 347 or 316, or molybdenum. Thin steel boats may be pressed or crimped and still be leaktight to molten contents; all others must be welded or specially fabricated. Because of their volatility, these metals must not be evacuated directly at higher temperatures, especially in the absence of the hydride phase. Several of the references on dissociation pressures describe closed containers permeable to hydrogen but not to metal vapor.

E. Measurement of Hydrogen Content of Sample

The amount of hydrogen taken up by the sample is determined by pressure–volume–temperature measurements before and after reaction. Included in the vacuum line is a glass bulb of known volume, which is standardized by weighing it filled with water before attachment to the line. This enables the calibration of the volumes of the remaining portions of the apparatus by systematic expansions from this bulb, which has been previously filled, to these remaining portions, which have been previously evacuated. Additional bulbs of other sizes, a mercury-filled gas buret, or both, might be added for additional range or accuracy.

It is necessary to consider a correction for hydrogen loss by diffusion through the walls of the stainless steel container. Such diffusion becomes appreciable above 650°C. It is determined directly in the apparatus by observed pressure decrease in a blank run. The author found 1.3×10^{-6} moles per minute per atmosphere$^{1/2}$ at 720°C. in a container of the size and type of Figure 3 (8).

If it is desirable to measure final uptake before the sample is cooled to room temperature, or to determine hydrogen uptake during the run, the effective "hot volume" of the sample container should be determined at a series of furnace temperatures. This is done by filling the calibrated system at room temperature to a pressure somewhat below 1 atm., observing manometer pressure versus furnace temperature, and calculating the volume of the system including the furnace as if the furnace were at room temperature.

For maximum accuracy, it is necessary to measure room temperature at several places along the vacuum line, as well as at the bulb, and to use an average room temperature weighted for the volumes of the various portions, to represent the temperature of the gas in the system.

Also, for maximum accuracy, it is necessary to keep the refrigerant level constant in the Dewar flask surrounding the cold trap. If solid CO_2 is used as refrigerant, a heavy chlorinated liquid such as trichloroethylene is used for heat transfer rather than acetone, to reduce temperature gradients in the trap.

A Toepler pump might be added to the line if it is desirable to quantitatively remove hydrogen from the sample, or to use the apparatus to analyze a portion of the sample for hydrogen content. The removal of the last portions of hydrogen from a sample is difficult, and, for the alkali and alkaline earth metals, is apt to be complicated by the volatility of these metals. However, if tin metal is added to most of these hydrides, an intermetallic compound of the metal with tin forms, which facilitates quantitative release of the hydrogen.

F. Analysis

Only methods which apply to all or most hydrides are discussed here. The principal method of assay is by hydrogen evolution by hydrolysis, followed by titration of the resulting solution. A sample weighing flask is fitted with a stopper or ground glass joint containing a separatory funnel, and a side outlet tube provided with a stopcock. Liquid is added through the funnel, and the side tube is attached to a system so that evolved hydrogen can displace water into a container for weighing.

The initial liquid to be added depends on the reactivity of the sample. It may be water, ice water, or pure, dry dioxane. The final added liquid is water if the hydroxide of the metal is soluble; dilute HCl if not. The total volume of added liquid must be measured. The HCl must be standardized, and the acid remaining in the funnel must be rinsed into the flask, if the resulting solution is to be titrated.

The metal content may be assayed by titration of the solution resulting from hydrolysis. Where the hydroxide is soluble, carbonate may be determined if the basic solution is first titrated to the Bromthymol Blue end point (pH 7), and then to the Methyl Orange end point (pH 4).

Alternatively, hydrogen may be evolved by vacuum decomposition using a Toepler pump (see earlier).

Two common impurities are nitride nitrogen and acetylide carbon. The former may be determined by the Kjeldahl method on the alkaline solution from hydrolysis. The carbon is best determined colorimetrically by the ammoniacal cuprous chloride method (9). On hydrolysis of the sample, acetylene is liberated with the hydrogen, and produces red cuprous acetylide which is determined colorimetrically.

III. LITHIUM HYDRIDE

A. Introduction

The preparation of lithium hydride follows the standard procedure closely. The pure hydride is a white to blue-gray coarsely crystalli1e solid. The blue color is due to small amounts of colloidal lithium precipitated on cooling from high temperatures where there may be hydrogen vacancies (10). It may be eliminated by scrupulous attention to purity and very slow cooling to assure maximum hydrogen absorption.

Lithium hydride is slowly photosensitive to visible light at room temperature, and rapidly to ultraviolet light and ionizing radiation. Other colors may be seen (11). The loss of hydrogen is usually too small to be detected analytically. The white color is restored by heating in hydrogen.

Less pure preparations may be more finely crystalline.

Pure lithium hydride melts at 688°C. under 1 atm. hydrogen pressure (8). It is the only hydride to melt without decomposition at 1 atm.

TABLE I
Lithium–Lithium Hydride Phase Equilibrium

	Mole fraction lithium hydride			
t,°C.	Metal phase, liquid	Hydride phase, solid	Hydride phase, liquid	Plateau, p_{mm}.[a]
625	0.13 (8)	0.99+	—	3.5 (12)
685	0.26 (8)	0.98–9 (8)	0.96–7 (13)	20 (13)
700	0.26 (13)	—	0.96 (13)	28 (13)
750	0.30 (13)	—	0.94 (13)	71 (13)

[a] $\log p_{mm.} = -9600/T + 11.227$ (12), 500–650°C. $\log p_{mm.} = -8224/T + 9.926$ (13), 700–800°C.

The two-phase system of metal plus hydride (Fig. 1) melts at the monotectic temperature of 685°C. (8), under about 20 mm. hydrogen pressure.

Table I gives the properties of the two-phase equilibrium mixture at several temperatures.

B. Kinetics of Formation of Lithium Hydride

The qualitative variation of rate of hydrogen uptake by bulk lithium with temperature has been observed to follow a definite pattern. It is readily observed and measured in an apparatus of the type described in Sec. II. The reaction begins at temperatures from 300–500°C., and proceeds at a moderate rate until a few per cent of theoretical hydrogen have been absorbed. The rate then slows down. Presumably the molten lithium phase has become saturated with hydrogen, and the rate becomes controlled by slow diffusion through the surface layer of the hydride phase. The temperature must be raised above the melting point of lithium hydride for a marked increase in rate, and completion of the reaction.

A sudden small *increase* in the hydrogen pressure may be noted at 300–350°C., followed by the decrease which indicates the beginning of reaction. This effect is due to the liberation of hydrogen in the reaction: $LiOH + LiH = Li_2O + H_2$. The LiOH impurity is able to react in this way at this temperature, which is presumably the LiH–LiOH eutectic. The magnitude of the effect is a qualitative measure of LiOH impurity, and decreases with decreasing contamination of the metal.

Lithium can be made to take up hydrogen at lower temperatures. The cleanliness of the surface is apparently very important. Remy-Genneté (14) reported that distilled lithium took up as much as 9% of theoretical hydrogen at room temperature in 24 hr. Hüttig and Krajewski (15) reported that Li dissolved in liquid ammonia reacted with hydrogen at room temperature. Li can also be made to react with hydrogen at lower temperatures under pressure in the form of a dispersion in an inert liquid (16,17), or in the presence of certain catalysts such as high molecular weight fatty acids or fused ring aromatic hydrocarbons (18).

Quantitative data on the rate of reaction are scarce, and to be questioned from the point of view of reproducibility, because of sensitivity to purity and surface condition of the metal. Albert and Mahé (19) observed a rate of hydrogen absorption of 1.5 cm.³ per cm.² per sec. at 1 atm. and 680°C. Soliman (20) reported that Li metal absorbed 118 times its own volume of hydrogen in 30 min. at 0.5 atm. and 500°C.

Swain and Heumann (21) studied the rate at temperatures from 25–250°C. They found the *initial* rate of decrease of hydrogen pressure to be proportional to the pressure and to the mass of lithium: $-dP/dT = kmp$. Actually the rate should be proportional to $p - p_e$, where p_e is the equilibrium pressure for the particular composition and temperature (Fig. 1). But p_e is very low in the Li–LiH system at these temperatures.

These authors give the following equation for the initial rate constant: $\log k = -1380/T + 2.09$. The heat of activation is 6.3 ± 2.5 kcal./ mole H_2, the variance being attributed to impurities. Later in the reaction, the rate constants decreased, the process apparently becoming diffusion controlled.

C. References to LiH Syntheses

An apparatus and procedure for the synthesis of LiH on the kilogram scale is described by Albert and Mahé (19). 99.6% pure LiH was made in 8 hr. An improved version of this apparatus is described by Ponomarenko and Mironov (22). Holding and Ross (23) describe a standard preparation of lithium deuteride.

Landa et al. (24) describe the preparation of LiH at 150–225 atm. and 310–345°C. 10 g. atoms of metal were reacted in a 2.5 l. stainless steel autoclave, with 0.1% WS_2 or MoS_2 as catalyst. Quantitative yields were reported.

Other references to the preparation of LiH, standard and nonstandard, are given by Messer (25).

D. A Procedure for the Preparation of LiH

This uses the apparatus and technique of Sec. II. The outer container or bomb is of stainless steel, and the sample boat is of stainless steel type 347 or 316, or of Armco iron. The $3^1/_2$-in. boat is suitable for the preparation of up to $^1/_2$ mole or 4 g. of LiH; boat, bomb, furnace, and volume calibration bulb must all be larger for larger preparations.

Foote Mineral Company lithium ingot, 99.96% Li, is satisfactory for direct use if a higher order of purity of final product is not needed. Otherwise, the metal must first be vacuum distilled (7).

The lithium metal is cut and trimmed of any oxide–nitride coating in an *argon*-filled drybox. (Carbon dioxide would be useable but less satisfactory; nitrogen definitely could *not* be used because it reacts rapidly with lithium at room temperature to form the dark nitride Li_3N.)

For highest purity of product, the sample boat must be pretreated with lithium hydride. A "dummy" synthesis of LiH is carried out, except that hydrogen pressure may be kept at 1 atm. throughout, with slow escape through the bubbler, and the sample is kept at 725°C. for 2 hr. instead of 8–10 hr. The boat is removed, dropped into pure methanol, and allowed to react until clean. It is washed with methanol and with anhydrous ether, transferred to the drybox lock wet with ether, and returned to the box for loading with the synthesis sample. This treatment seems to make the surface of the boat more impervious to attack by lith-ium metal.

The desired amount of lithium, up to 4 g. sample size, is cut, in the dry-box, into pieces which fit easily into the boat. A heavy, sharp knife may be used. The boat is loaded into the outer container which is closed, re-moved from the drybox, attached to the line, and evacuated. The lith-ium sample in the bomb is now heated to 200°C. for about $^{1}/_{2}$ hr. to de-gas the bomb and the sample. The valve to the bomb is closed, and the rest of the calibrated system filled with hydrogen. After P–V–T meas-urements of this hydrogen, the valve is opened, exposing the sample to hydrogen.

The temperature is raised to 725°C., and kept constant at this tem-perature until the reaction is complete. Each time the pressure drops to 100 mm. or so, the valve to the sample bomb is closed, P–V–T meas-urements are made, and additional measured hydrogen is admitted to 1 atm. P, V, and T are read, and the valve to the bomb is opened. This process is continued until the rate of absorption is too slow to detect. The final pressure must be at or just below 1 atm. for highest hydrogen content.

Exposure for 2–10 hr. at 720–725°C., followed by the slow cooling process, was found sufficient to prepare a sample containing 99.6–99.8% of theoretical hydrogen. The variation is probably due to the purity of the metal, since the preparations with the 99.96% metal required the shorter times.

The sample is cooled very slowly to 680°C. to allow maximum uptake during the freezing process. The sample is further cooled in steps of 100°C., down to 300°C., as long as significant hydrogen absorption is noted. Final P–V–T measurements are made either at the last annealing temperature or at room temperature.

The sample is unloaded in the drybox. The bulk of the crystals may be removed from the small stainless steel boat by bending the boat; from a larger boat they may be removed by tapping. An iron boat is more brittle and the sample must be removed with care. Any off-color por-tions of the preparation should be discarded.

E. Analysis

Lithium hydride may be analyzed by the methods of Sec. II. Ice water may be used to start the hydrolysis.

Friedman (26) describes methods and references for trace analysis of LiH for several elements. Bergstresser and Waterbury (27) describe the evolution of hydrogen from LiH with lead metal at 600°C.

IV. SODIUM HYDRIDE

A. Introduction

The synthesis of sodium hydride from the elements cannot be carried out by the standard procedure because of its high melting point and high dissociation pressure. These make the rate of the reaction between hydrogen and bulk metal impractically slow at the low temperatures at which sodium hydride can exist at 1 atm. pressure.

By analogy with the known melting points of LiH, LiF, LiCl, NaF, and NaCl, the melting point of NaH may be estimated to be of the order of magnitude of 860°C.

Dissociation pressures and isotherm data for the sodium–hydrogen system are given in Table II.

TABLE II
Sodium–Sodium Hydride–Hydrogen Equilibrium

| | Plateau[a] | | | 100%[b] |
t,°C.	% NaH, metal phase	% NaH, hydride phase	Pressure, p_{mm}.	Pressure, p_{mm}.
350	—	—	74 (28)	220 (29)
422	—	—	760 (28,29)	1500 (29)
550	30 (29)	80 (29)	16600 (29)	21400 (29)

[a] Plateau: $\log p_{mm}. = -6100/T + 11.66$ (28); $\log p_{mm}. = -5958/T + 11.47$ (29).

[b] "100%": $\log p_{mm}. = -5070/T + 9.49$ (29).

Thus the sodium hydride phase cannot exist above 422°C. at 1 atm. hydrogen pressure. In the purest, near-stoichiometric "100%" form, which has a higher dissociation pressure, sodium hydride cannot exist above about 390°–400°C., according to these data.

It should be noted that, although the truly "flat" portions of the isotherms of Banus et al. (29) run from approximately 30–80% NaH, the approximately flat portions run from about 5–85% NaH. These represent estimates of the mutual solubility of Na and NaH at temperatures of about 500–600°C. Williams et al. (30) measured the solubility of sodium hydride in sodium metal at the plateau pressure as a function of temperature, and obtained values of 0.3 wt. % at 350°C. and 4.25% at 445°C.

At these low temperatures, although the first hydrogen may dissolve in and react with the sodium fairly rapidly, the coating of hydride which soon forms on the surface of the metal becomes virtually impermeable to hydrogen, causing absorption to cease. Gibb (2) suggests that this is partly due to the contraction of the metal on hydride formation.

Keyes (31), as part of his early study of the dissociation pressures of sodium and potassium hydrides, gives a small amount of semiquantitative data on the initial rate of absorption of hydrogen by metal, in terms of the pressure drop noted in a given length of time. He noted that the reaction is 16 times faster at 300°C. than at 200°C., and immeasurably fast above 300°C.

The reaction between the elements to form sodium hydride is effected quantitatively at a reasonable rate: (a) by the use of finely dispersed metal, or (b) by the use of a very small amount of a reagent which will break up the liquid metal–solid hydride interface, or (c) by condensation from the vapor phase.

B. The Dispersion Method

This procedure for sodium hydride is given in full detail for laboratory duplication by Mattson and Whaley in *Inorganic Syntheses* (32). A sodium metal dispersion is made by stirring sodium with heavy white mineral oil containing a trace of oleic acid, under nitrogen at 105–125°C. In the same apparatus, the temperature is raised and the feed gas is changed from nitrogen to hydrogen. After final treatment at 280–300°C., the sample is cooled, and the feed gas changed back to nitrogen. The sodium hydride is recovered from the dispersion by washing with hexane and decantation. The yield is quantitative.

C. The Use of a Catalyst or Surface Active Agent

In the presence of any one of a considerable number of substances, the direct synthesis of sodium hydride will proceed to completion. The amount needed varies from 0.1% to several per cent. Often the hydride is pure enough for the purpose at hand without its removal; if not it must be removed by hexane extraction or otherwise.

Hansley (18,33) found that $0.1–1.0\%$ of any of the following facilitated rapid absorption of hydrogen at 250–400°C. and 1 atm.: fatty acids with 8 or more carbon atoms, metal salts of these acids, anthracene, fluorene, indene, and various isopropyl-substituted aromatic hydrocarbons.

Muckenfuss (16) synthesized sodium hydride using the metal dispersed in tetralin or Nujol, at 250°C. and 500 p.s.i. in a rocking autoclave. This procedure was adapted by Sayre and Beaver (34), who describe their synthesis in sufficient detail for repetition.

The autoclave used by these authors was an Aminco 406–22B stainless steel bomb. It was provided with a 600 ml. Pyrex liner.

The surface active agent was a petroleum cut boiling in the 167–180°C. range.   The authors considered this range quite critical.   Lower boiling solvents did not work as well, presumably due to their lower viscosity. Higher boiling solvents would have required washing the preparation with more volatile solvents.

A single piece of sodium (about $^1/_2$ mole) was cut under decane, and transferred to the autoclave under the surface-active solvent.   The sodium was hydrided in the rocking autoclave at 240°C. and 400 p.s.i. for 10 hr.   About 300 ml. solvent was used.

The resulting hydride slurry was transferred under the liquid to a special drying chamber described by the authors.   This chamber contained glass balls so that the contents could be agitated while the solvent was pumped off under high vacuum.   This facilitated the breaking up of the packed hydride powder and more complete solvent removal.

Analytical data by hydrogen evolution and titration of base agreed well, showing high purity of the sample.

Landa et al. (24) describe the preparation of sodium hydride by the same method used for lithium hydride (Sec. III. C).   10 moles of sodium were reacted with hydrogen at 130–210 atm. and 220–250°C., using 0.1% WS_2 as catalyst.

Mikheeva et al. (35) give a very thorough discussion of the synthesis of sodium hydride, several procedures with diagrams, and a comparison of catalysts with discussion of the catalytic process.

These authors tried three types of procedures: the static combination of sodium and hydrogen, the circulation of hydrogen over or the bubbling of hydrogen through molten sodium, and the mechanical mixing of the reagents.   They found very little hydride formed at 200–400°C. with the static method, even in the presence of palmitic or stearic acid or a salt of one of these acids.   Circulation of hydrogen above the sodium improved the yield only somewhat.   Bubbling the hydrogen through a suspension of sodium in o-xylene or kerosene with 3–6% anthracene improved the yield to 76% of a black product.

However, the use of an autoclave, and stirring with a perforated rectangular plate at 300–600 r.p.m., improved the yield in the best runs to 96–97%, and gave a light-gray product.   The pressure was about 0.7 atm., the temperature 270°C., the time of hydriding 2.5 hr., and the amount of catalyst 0.25–0.5% in these best runs.

"Industrial Oil 30" was found to give a better yield than anthracene in these runs, and the "benzene aromatic fractions" of this oil gave the highest yields.

These authors give an interesting hypothesis for the catalytic action of anthracene and related materials on this reaction.   They point out that the phenomena associated with the reaction of sodium metal with anthracene and other polycyclic hydrocarbons are similar to the phenomena noted in the hydriding of sodium in the presence of these compounds. In the former case, sodium reacts with anthracene to form the 9,10 disodium salt of 9,10 dihydroanthracene (36). In the latter case, it is suggested that anthracene thus plays the role of a sodium carrier, reacting first with the metal to form the 9,10 salt, which then reacts with hydrogen to give sodium hydride and anthracene. Other aromatic surface active agents would be those that could react with sodium in the same way.

Support for this mechanism is claimed from the researches of Hugel (37,38) who offered the hypothesis that sodium hydride would catalyze the hydrogenation of aromatic hydrocarbons only at positions in the molecule where sodium metal itself could add.

D. Vapor Phase Reaction

Sodium hydride may be synthesized by the action of the elements in the vapor phase. Hydrogen gas is passed over or through the metal, and sodium hydride condenses out as a fine crystalline mass on the walls of the colder parts of the apparatus. Because of its slowness, necessary to avoid the deposition of metallic sodium, this method is suited only for the preparation of small amounts of hydride.

Bardwell (39) prepared both sodium and potassium hydrides by bubbling hydrogen through molten metal in a steel tube at 400°C. Fine white needles condensed in the cold parts of the apparatus. The apparatus, which appears quite simple, is illustrated. The quantities, not stated, appear to be of the order of 1 g. The vapor pressure of sodium metal at this temperature is just under 1 mm. Hg (40).

Hagen and Sieverts (41) heated technical sodium hydride in an iron crucible in a glass apparatus to 420°C. for 4–8 hr., and obtained up to 30 mg. of fine white needles of NaH.

The author and co-workers (42) used this method for the preparation of sodium hydride in quantities of several grams. Hydrogen was passed over molten sodium for 24–48 hr. at 1 atm. and 650–750°C. Here the vapor pressure of sodium is of the order of 100 mm. In general, the hydride condensed in a cottony mass, although in one or two cases needle-like crystals were observed. Some rejected gray portions were formed along with the white. The preparations seemed quite sensitive to residual drybox moisture, and had to be handled fairly rapidly.

This method, although suited only with great difficulty for the preparation of sizeable amounts of hydride, does give pure preparations, and deserves a full investigation, with attention to such factors as temperature, rate of passage of hydrogen, temperature gradients, and position of the condensation surface, etc.

V. OTHER ALKALI METAL HYDRIDES

Potassium hydride is very similar to sodium hydride, so that the conditions of its preparation are nearly the same. The dissociation pressure of potassium hydride is very slightly lower than that of sodium hydride (28,31). Its rate of formation, under fairly comparable conditions, was faster than that of sodium hydride (24).

Mikheeva and Shkrabkina (43) describe the preparation of KH in an autoclave at 200–250°C. and 3–4 atm. hydrogen pressure in the presence of 1% machine oil. The yield was 97–98% KH.

Rubidium and cesium hydrides are less well known. Their dissociation pressures are higher than those of sodium or potassium hydrides (28). Presumably the trend toward increasing rate of formation continues.

VI. CALCIUM HYDRIDE, CaH₂

A. Introduction

Calcium hydride, at least up to 99% of theoretical hydrogen, may be prepared by the standard procedure. It is a white powder or finely crystalline solid, with melting point well above 1000°C. (44). Under "plateau" conditions (Sec. I) it is more stable than lithium hydride, but at near-stoichiometric compositions it is less stable. Thus the addition of the last traces of hydrogen is difficult. The mutual solubilities of metal and hydride phases are greater than for Li and LiH at temperatures from 500–800°C.

The solid–liquid phase diagram of the Ca–CaH₂ system, by thermal analysis and equilibrium experiments, was obtained by Peterson and Fattore (44). The system is complicated by the existence of three allotropic forms of the calcium phase and two of the hydride phase in various temperature regions.

Dissociation pressure isotherms of the type of Figure 1, and plateau dissociation pressures have been measured by a considerable number of workers. The three most recent researches—those of Johnson and co-workers (45), Treadwell and Sticher (46), and Curtis and Chiotti (47)—are cited here; these refer to earlier references.

Table III gives dissociation pressure and solubility results from these references. There are minor discrepancies, but good qualitative agreement. The plateau dissociation pressure is small below 600°C., and the solubility of hydrogen in calcium is small below 400°C. At 800–900°C., the solubility of CaH_2 in Ca is about 20%; the hydride phase at equilibrium is about 95% CaH_2.

TABLE III
Calcium–Calcium Hydride–Hydrogen Equilibrium

	Plateau[a]			
$t,°$ C.	Mole fraction CaH_2, metal phase (44)	Mole fraction CaH_2, hydride phase (44,46)	Pressure, p_{mm}. (47)	99.14% CaH_2 pressure, p_{mm}. (46)
400	0.02	—	—	—
500	0.05	—	—	—
600	0.07	~0.95	0.2	—
700	0.11	~0.95	2.2	10
800	0.20	~0.95	19	70
900	0.22	~0.95	94	550

[a] Plateau: $\log p_{mm}. = -10870/T + 11.49$, 650–900°C. (45); $-9840/T + 10.7$ 711–941°C. (46); $-9610/T + 10.23$, 578–780°C. (47); $-8890/T + 9.54$, 780–900°C. (47).

B. Kinetics of Formation of Calcium Hydride

The rate of combination of calcium metal with hydrogen has been studied by a number of workers with significant results. However, sensitivity to residual impurities seems to be the limiting factor in quantitative reproduction of rate constants from sample to sample.

Johnson et al. (45) studied the reaction from 0–200°C., using finely divided metal prepared from the evaporation of a solution of calcium metal in liquid ammonia. For temperatures above 100°C. and pressures above 0.05 atm. the results were irregular. From 27–98°C., at pressures below 15 mm., the measurements were consistent. They indicated a first-order reaction (log p_{H2} vs. time was linear), with a heat of activation of 7.7 ± 0.5 kcal./mole.

Treadwell and Sticher (46) similarly found a first-order reaction at 327–356°C., 8–130 mm., with heat of activation 7.8 kcal., using bulk metal.

Shushunov and Shafiev (48) studied the reaction from 125–500°C. using a film of metal distilled onto a glass surface. They found *two* sets of straight lines in Arrhenius plots of the rates: one of smaller slope at higher temperatures and lower pressures, with heat of activation 5.5–7.2 kcal., and one of larger slope at lower temperatures and higher pressures, heat of activation 12–15 kcal. First-order kinetics was associated only with the lower heats of activation; it was concluded that the higher values indicated that the reaction was controlled by diffusion through the calcium hydride layer under these conditions.

Kawana (49–51) has measured and calculated the rates of absorption of hydrogen and deuterium by calcium from 250–325°C. He considered the surface and nucleation aspects. He found the process to be first order, and obtained heats of activation of 13.8 and 13.9 kcal./mole, respectively.

Soliman (20) found: (a) a maximum in the rate of the reaction at 350°C., and (b) a marked increase in the rate of absorption with the use of pressures above atmospheric.

Shushunov and Shafiev (48) measured the effect of deliberate introduction of oxygen and water vapor impurities on the rate of the reaction. 0.5% O_2 reduced the rate to 42% of its original value; 1% or more reduced the rate to 1–3% of the original. The results for H_2O were very similar, the corresponding values being 15% and 1.5–2% of original, respectively.

C. The Preparation of Calcium Hydride

Johnson et al. (45) were able to prepare 99–100% CaH_2 at 230–270°C., but could not make preparations better than 91% at 650–900°C.

Treadwell and Sticher (46) prepared 99.14% CaH_2 at 380°C. in a version of the standard apparatus, under static conditions. However, they found they could not exceed this hydrogen content without the use of stirring. They describe an apparatus in which this could be done. The hydriding was done to the extent of 96% at 500°C. without stirring. The apparatus could then be turned so that a magnetic stirrer could be introduced. The hydriding was then continued at 500°C., in all cases additional hydrogen being taken up. The best preparations gave 99.6–99.9% CaH_2.

All of these workers vacuum distilled their calcium metal before hydriding, usually at 10^{-3} to 10^{-4} mm. at 850–900°C. Johnson (45) and Treadwell (46) describe their stills.

A source of vacuum distilled calcium is Dominion Magnesium, Limited, of Toronto, Canada.

The need for vacuum distillation is based upon a comparison of the purity of the original with the purity required in the desired product.

The standard apparatus and technique may be used, as for lithium hydride (Sec. III). As much of the reaction as possible should be carried out at 300–350°C.   If it is necessary or desirable to heat the sample to a higher temperature, it should be kept at 300–350°C. on cooling until no further hydrogen is taken up over an hour or longer.

If the calcium hydride sample is to be heated above 600°C., an all iron or stainless steel apparatus is recommended.   In any event, the sample boat should always be of iron or stainless steel.

Calcium hydride may be prepared by condensation from the vapor mixture of calcium and hydrogen.   Zintl and Harder (52), in connection with their x-ray crystallographic studies of the hydrides of calcium, strontium, and barium, prepared crystals of these substances in this way, occasionally up to 1 mm. in size.

VII. OTHER SALINE HYDRIDES

A. Introduction

These include the dihydrides of Ba, Sr, Eu, Yb, and Mg.   All except MgH_2 are very similar to CaH_2, physically and crystallographically. EuH_2 and YbH_2 have the same orthorhombic crystal structure (53) as CaH_2, SrH_2, and BaH_2 (52).   This results from the peculiar stability of the $2+$ oxidation state of these metals.

B. Barium and Strontium Hydrides

These have not been studied as thoroughly as has calcium hydride. Peterson and Indig (54) have measured the solid–liquid phase diagram of the Ba–BaH_2 system.   Gibb (55) gives dissociation pressure data for the Sr–SrH_2 and Ba–BaH_2 systems.   The following empirical equations represent what are presumably the plateau pressures:

$$\text{Sr–}SrH_2\text{:} \ \log p_{mm.} = -10400/T + 11.10 \ (92.3\% \ SrH_2)$$

$$\text{Ba–}BaH_2\text{:} \ \log p_{mm.} = -6450/T + 8.20 \ (93.7\% \ BaH_2)$$

The phase diagram indicates higher miscibility in the Ba–BaH_2 system than in the Ca–CaH_2 system.   According to the dissociation pressure data, the stabilities of SrH_2 and CaH_2 are about the same under plateau conditions; BaH_2 is definitely less stable than the other two.

C. Magnesium Hydride

MgH_2 has the rutile structure (56), rather than the orthorhombic structure of CaH_2, SrH_2, and BaH_2.   It is also much less stable than these other hydrides, reaching a plateau dissociation pressure of 1 atm. at 280–290°C. (56–58).

The rate of formation of MgH_2 (56–58) is such that it is necessary to use pressures above 1 atm. and temperatures above 290°C. to synthesize the compound from the elements in a reasonable time. In the range of pressures from 100–1000 mm., and temperatures from 250–300°C., days were required for equilibrium (58).

Wiberg et al. (59) prepared MgH_2 from the elements at 570°C. and 200 atm., but in only 60% yield, and with MgI_2 as catalyst. Dymova et al. (60) used a rotating autoclave and CCl_4 catalyst, at 100–200 atm. and 310–450°C. Faust et al. (61) used allyl iodide, propargyl iodide, and iodine as catalysts, at 5 atm. and 175°C. However, the dissociation pressure results seem to indicate that a catalyst is not really necessary at and above 400°C.

Those workers reporting analyses of product indicate that it is extremely difficult to prepare magnesium hydride without considerable amounts of oxide and metal impurity. Stampfer et al. (57) report 93.1% MgH_2, 3.0% MgO, and 3.9% Mg.

MgH_2 may also be prepared by thermal decomposition of magnesium dialkyls (62,63).

REFERENCES

1. Hurd, D. T., *Chemistry of the Hydrides*, Wiley, New York, 1952.
2. Gibb, T. R. P., Jr., "Primary Solid Hydrides," in *Progress in Inorganic Chemistry*, Vol. III, F. A. Cotton, ed., Interscience, New York, 1962, pp. 315–509.
3. Pretzel, F. E., G. N. Rupert, C. L. Mader, E. K. Stearns, G. V. Gritton, and C. C. Rushing, *Phys. Chem. Solids* 16, 10 (1960).
4. Proskurnin, M., and J. Kasarnowsky, *Z. Anorg. Allgem. Chem.* 170, 301 (1928).
5. Zintl, E., and A. Harder, *Z. Physik. Chem. (Leipzig)* B14, 265 (1931).
6. Gibb, T. R. P., Jr., *Anal. Chem.* 29, 584 (1957).
7. Douglas, T. B., L. F. Epstein, J. L. Dever, and W. H. Howland, *J. Am. Chem. Soc.* 77, 2144 (1955).
8. Messer, C. E., E. B. Damon, P. C. Maybury, J. Mellor, and R. A. Seales, *J. Phys. Chem.* 62, 220 (1958).
9. Weaver, E. R., *J. Am. Chem. Soc.* 38, 352 (1916).
10. Pretzel, F. E., and C. C. Rushing, *Phys. Chem. Solids* 17, 232 (1961).
11. Moers, K., *Z. Anorg. Allgem. Chem.* 113, 179 (1920).
12. Hurd, C. B., and G. A. Moore, *J. Am. Chem. Soc.* 57, 332 (1935).
13. Heumann, F. K., and O. N. Salmon, "The Lithium Hydride, Deuteride, and Tritide Systems," KAPL-1667, Dec. 1, 1956; *Nucl. Sci. Abstr.* 11, 4811 (1957).
14. Remy-Genneté, P. A., *Ann. Chim.* 19, 263 (1933).
15. Hüttig, G. F., and A. Krajewski, *Z. Anorg. Allgem. Chem.* 141, 133 (1924).
16. Muckenfuss, A. M., U.S. Patent 1,958,012, May 8, 1934; through *Chem. Abstr.* 28, 4185 (1934).
17. The Roessler and Hasslacher Chemical Co., British Patent 405,017, Jan. 29, 1934; through *Chem. Abstr.* 28, 4545 (1934).
18. Hansley, V. L., U.S. Patents 2,372,670 and 2,372,671, April 3, 1945; through *Chem. Abstr.* 39, 3129 (1945).

19. Albert, P., and J. Mahé, *Bull. Soc. Chim. France* **1950**, 1165.
20. Soliman, A., *J. Appl. Chem.* (*London*) **1**, 98 (1951).
21. Swain, E. E., and F. K. Heumann, "The Reaction between Lithium and Hydrogen at Temperatures between 29°C. and 250°C.," KAPL-1067, March 1, 1954: *Nucl. Sci. Abstr.* **9**, 5589 (1955).
22. Ponomarenko, V. A., and V. P. Mironov, *Izvest. Akad. Nauk SSSR, Otd. Khim. Nauk* **1954**, 407; *Bull. Acad. Sci. USSR Div. Chem. Sci.* **1954**, 423.
23. Holding, A. F. I., and W. A. Ross, *J. Appl. Chem.* (*London*) **8**, 321 (1958).
24. Landa, S., F. Petru, J. Mostecky, J. Vit, and V. Prochazka, *Collection Czech. Chem. Commun.* **24**, 2037 (1959).
25. Messer, C. E., NYO-9470, Oct. 27, 1960; *Nucl. Sci. Abstr.* **15**, 12906 (1961).
26. Friedman, H. A., *Anal. Chem.* **32**, 137 (1960).
27. Bergstresser, K. S., and G. R. Waterbury, "Determination of Hydrogen in LiH," LAMS-1698, May, 1943, decl. Oct. 5, 1956; *Nucl. Sci. Abstr.* **13**, 1116 (1959).
28. Herold, A., *Compt. Rend.* **228**, 686 (1949).
29. Banus, M. D., J. J. McSharry, and E. A. Sullivan, *J. Am. Chem. Soc.* **77**, 2007 (1955).
30. Williams, D. D., J. A. Grand, and R. R. Miller, *J. Phys. Chem.* **61**, 379 (1957).
31. Keyes, F. G., *J. Am. Chem. Soc.* **34**, 779 (1912).
32. Mattson, G. W., and T. P. Whaley, "Sodium Hydride," in *Inorganic Syntheses*, Vol. 5, T. Moeller, ed., McGraw-Hill, New York, 1957, pp. 10–13.
33. Hansley, V. L., and P. J. Carlisle, *Chem. Eng. News* **23**, 1332 (1945).
34. Sayre, E. V., and J. J. Beaver, *J. Chem. Phys.* **18**, 584 (1950).
35. Mikheeva, V. I., T. N. Dymova, and M. M. Shkrabkina, *Zh. Neorgan. Khim.* **4**, 709 (1959); *Russian J. Inorg. Chem.* (*English Transl.*) **4**, 323 (1959).
36. Schlenk, W., J. Appenrodt, A. Michael, and A. Thal, *Chem. Ber.* **47**, 473 (1914).
37. Hugel, G., and J. Friess, *Bull. Soc. Chim. France* **49**, 1042 (1931); through *Chem. Abstr.* **26**, 363 (1932).
38. Hugel, G., and Gidaly, *Bull. Soc. Chim. France* **51**, 639 (1932); through *Chem. Abstr.* **26**, 5002 (1932).
39. Bardwell, D. C., *J. Am. Chem. Soc.* **44**, 2499 (1922).
40. Makansi, M. M., C. M. Muendel, and W. A. Selke, *J. Phys. Chem.* **59**, 40 (1955).
41. Hagen, H., and A. Sieverts, *Z. Anorg. Allgem. Chem.* **185**, 239 (1930).
42. Messer, C. E., L. G. Fasolino, and C. E. Thalmayer, *J. Am. Chem. Soc.* **77**, 4524 (1955).
43. Mikheeva, V. I., and M. M. Shkrabkina, *Zh. Neorgan. Khim.* **7**, 463 (1962); through *Chem. Abstr.* **57**, 373 (1962).
44. Peterson, D. T., and V. G. Fattore, *J. Phys. Chem.* **65**, 2062 (1961).
45. Johnson, W. C., M. F. Stubbs, A. E. Sidwell, and A. Pechukas, *J. Am. Chem. Soc.* **61**, 318 (1939).
46. Treadwell, W. D., and J. Sticher, *Helv. Chim. Acta* **36**, 1820 (1953).
47. Curtis, R. W., and P. Chiotti, *J. Phys. Chem.* **67**, 1061 (1963).
48. Shushunov, V. A., and A. I. Shafiev, *Zh. Fiz. Khim.* **26**, 672 (1952); AEC-translation-3220, *Nucl. Sci. Abstr.* **12**, 9666 (1958).
49. Kawana, Y., *Nippon Kagaku Zasshi* **71**, 494 (1950); through *Chem. Abstr.* **45**, 6468 (1951).
50. Kawana, Y., *Nippon Kagaku Zasshi* **71**, 554 (1950); through *Chem. Abstr.* **45**, 6468 (1951).
51. Higuchi, I., and Y. Kawana, *Nippon Kagaku Zasshi* **71**, 624 (1950); through *Chem. Abstr.* **45**, 7461 (1951).

52. Zintl, E., and A. Harder, *Z. Elektrochem.* **41,** 33 (1935).

53. Warf, J. C., and W. L. Korst, *Acta Cryst.* **9,** 452 (1956).

54. Peterson, D. T., and M. Indig, *J. Am. Chem. Soc.* **82,** 5645 (1960).

55. Gibb, T. R. P., Jr., "Hydrides and Metal-Hydrogen Systems. Final Report," *NEPA*-**1841**, April 30, 1951, declass. July 18, 1961; *Nucl. Sci. Abstr.* **16,** 8713 (1962).

56. Ellinger, F. H., C. E. Holley, Jr., B. B. McInteer, D. Pavone, R. M. Potter, E. Staritzky, and W. H. Zachariasen, *J. Am. Chem. Soc.* **77,** 2647 (1955).

57. Stampfer, J. F., Jr., C. E. Holley, Jr., and J. F. Suttle, *J. Am. Chem. Soc.* **82,** 3504 (1960).

58. Kennelley, J. A., J. W. Varnig, and H. W. Myers, *J. Phys. Chem.* **64,** 703 (1960).

59. Wiberg, E., H. Goeltzer, and R. Bauer, *Z. Naturforsch.* **6b,** 394 (1951).

60. Dymova, T. N., Z. K. Sterlyadkina, and V. G. Safronov, *Zh. Neorgan. Khim.* **6,** 763 (1961); *Russian J. Inorg. Chem. (English Transl.)* **6,** 389 (1961).

61. Faust, J. P., E. D. Whitney, H. D. Batha, T. L. Heyring, and C. E. Fogle, *J. Appl. Chem.* **10,** 187 (1960).

62. Wiberg, E., and R. Bauer, *Chem. Ber.* **85,** 593 (1952).

63. Freundlich, W., and B. Claudel, *Bull. Soc. Chim. France* **1956,** 967.

CHAPTER 9

Sulfur–Nitrogen–Fluorine Compounds

OSKAR GLEMSER

Institute of Inorganic Chemistry, University of Göttingen, Germany

CONTENTS

I. INTRODUCTION

Sulfur–nitrogen–fluorine compounds were first prepared in 1955 (1). Since then new fluorides of this type have been found. Structural studies have shown that they may be divided into two categories: acyclic and cyclic compounds.

Acyclic Compounds. Thiazylfluoride, NSF; thiodithiazyldifluoride, $S_3N_2F_2$; thiazyltrifluoride, NSF_3.

Cyclic Compounds. Tetrathiazyltetrafluoride, $N_4S_4F_4$; trithiazyltrifluoride, $N_3S_3F_3$; thiotrithiazylfluoride, S_4N_3F.

All these fluorides can be prepared with tetrasulfur tetranitride, S_4N_4, as starting material. The chemical bonding in S_4N_4 and its derivatives, the acyclic and cyclic sulfur–nitrogen–fluorine compounds, constitutes a rather unusual situation, unique in the field of pure inorganic chemistry. There are S–N single, double, and triple bonds, as well as localized and delocalized π-bonds in rings as shown by the structural formulas in Figure 1.

As in all fluorination processes, side reactions with the container material may become important and, therefore, need special attention. The commonly used container materials are silica, copper, polyethylene,

227

and Teflon. All the above fluorides, with the exception of $N_4S_4F_4$ and mainly NSF_3, are extraordinarily sensitive to moisture. This fact must be taken into consideration during their preparation.

Whereas a detailed concept of the stepwise hydrolysis of some of these compounds has been worked out, nothing is known as yet about the reac-

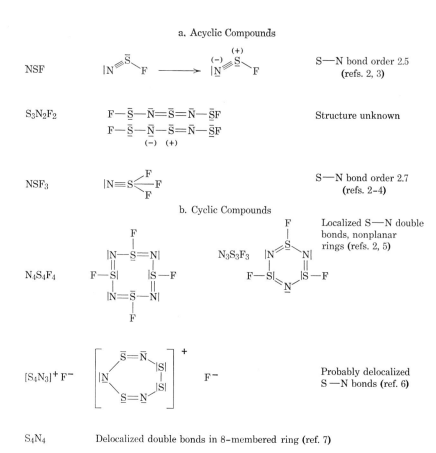

Fig. 1. Structural formulas of sulfur–nitrogen–fluorine compounds.

tion mechanism involved during their formation. The only thing known is that very frequently the reactions of tetrasulfur tetranitride, S_4N_4, suggest the existence of the intermediate radical SN. The fluorination of S_4N_4 to the gaseous monomer NSF bears out this hypothesis. However, the formation of $N_4S_4F_4$ cannot be explained in this way, as will be further discussed in Sec. III.A.1.

II. ACYCLIC COMPOUNDS

A. Thiazylfluoride, NSF

1. Preparation

Thiazylfluoride, NSF, can be prepared in various ways. Passing ammonia into a suspension of sulfur and silver difluoride in carbon tetrachloride yields NSF (8) (and NSF_3, see Sec. II.C) according to the equation

$$NH_3 + S + 4AgF_2 \xrightarrow{\text{CCl}_4} NSF + 3HF + 4AgF$$

Also, ammonia can react with excess SF_4 to give NSF (9) according to the equation

$$NH_3 + SF_4 \rightarrow NSF + 3HF$$

The fluorination of S_4N_4 is a suitable method for preparing NSF. The best results are obtained by the fluorination of a suspension of S_4N_4 in boiling carbon tetrachloride by means of AgF_2 (10) or HgF_2 (11). Using AgF_2, numerous other compounds are obtained besides NSF. Their separation from the reaction mixture is difficult and tedious. Using HgF_2, the raw product is relatively pure. The preparation of NSF with the latter fluorinating agent is more convenient and gives the highest yields.

$$S_4N_4 + 4HgF_2 \xrightarrow{\text{CCl}_4} 4NSF + 2Hg_2F_2$$

Procedure. NSF (11) has been prepared in a cylindrical silica vessel with reflux condenser to which two cold traps are attached; 3 g. S_4N_4 and 100 ml. of freshly distilled water-free carbon tetrachloride are put into the silica vessel. From a copper tube we add about 80 g. HgF_2 which has been freshly prepared in the same tube by heating 100 g. $HgCl_2$ in fluorine gas to $100\,°C.$, and subsequently cooling it in a stream of fluorine. Immediately after the addition of the HgF_2, the suspension is heated to the boiling point of the carbon tetrachloride, while being agitated vigorously with a magnetic stirrer. A very slow stream of nitrogen carries the volatile products through the reflux condenser into a trap cooled to $-80\,°C.$ The following trap (cooled to $-183\,°C.$) serves to exclude rediffusion of moisture. After 5–6 hr. the fluorination is complete.

The crude product is purified by fractional condensation in high vacuum. After removing the cold bath from the trap in which the crude product has been collected, the volatile components are passed through three cold traps. In the first trap ($-80\,°C.$) carbon tetrachloride condenses, in the second (−115°C.) NSF condenses, while in the third (cooled with liquid air) the highly volatile impurities such as SOF_2, SF_4, and CCl_3F are collected. Further purification of the NSF may be achieved by repetition of the fractional condensation.

Yield. 1.5–2 g. NSF which corresponds to 35–47% of the theoretical value as calculated from the amount of S_4N_4 consumed.

2. *Properties*

NSF is a colorless, pungently smelling gas with a boiling point of 0.4°C., as taken from the vapor pressure curve. The freezing point is −89°C. The "chemical shift" of the fluorine nuclear magnetic resonance signal (using saturated aqueous potassium fluoride solution as a reference) lies at +358 p.p.m. (2). The infrared spectrum shows absorption bands at the fundamental frequencies 1372, 640, and 360 cm.$^{-1}$.

NSF is highly reactive; the gas undergoes hydrolysis with water vapor yielding, as an intermediate product, thionyl imide HNSO which is readily recognized by its infrared spectrum (10). The end products of the quantitative hydrolysis with dilute HCl are fluoride, sulfite, and ammonium ions. In a copper vessel, NSF can be stored at room temperature and at a pressure of about 500 mm. Hg for several days without decomposition. In glass under similar conditions, decomposition takes place. S_4N_4 and greenish-yellow crystals of $S_3N_2F_2$ are formed on the glass walls. After heating the glass container to 110°C. for 3 hr., all the NSF is decomposed and the compounds S_4N_4, SiF_4, SO_2, SOF_2, and N_2 are formed. NSF can, in analogy to NSCl (12), polymerize to form $N_3S_3F_3$ (see Sec. III.B.1).

B. **Thiodithiazyldifluoride, $S_3N_2F_2$ (13)**

1. *Preparation*

If one fills a 6-l. glass flask with NSF at room temperature and at an initial pressure of about 600 mm. Hg after 1 hr. a yellow coating starts to appear on the inside walls. Later, greenish-yellow crystalline platelets start to grow. After 1 week, the greenish-yellow crystals can be scraped off with a needle controlled from the outside by means of a magnet; these crystals are separated from the simultaneously formed yellow coating. They are collected in a multichamber sublimation tube attached to the bottom of the 6-l. flask. By high vacuum sublimation, one obtains, at 40°C. yellowish green platelets, and at 65°C. bright-green pointed crystals. Both fractions dissolve in carbon tetrachloride, yielding the same yellowish green solution and giving identical ultraviolet absorption spectra with a broad maximum at about 375 mμ. Presumably these fractions are polymorphic modifications of the same compound. Thirty runs yield about 300 mg. of this compound while 300 g. of S_4N_4 are consumed—corresponding to a yield of approximately 0.1%.

2. Properties

$S_3N_2F_2$ melts at 83°C. Between 85–97°C., decomposition starts; above 100°C. $S_3N_2F_2$ explodes. Contact with moisture causes the fluoride to turn immediately black. By alkaline hydrolysis the nitrogen is quantitatively released in the form of ammonia.

C. Thiazyltrifluoride

1. Preparation

As mentioned in Sec. II.A.1, NSF_3 (as well as NSF) can be prepared by passing ammonia into a suspension of sulfur and silver difluoride in carbon tetrachloride (8). Nothing is known about the mechanism of the formation. Hence, one can only formulate the reaction as follows:

$$NH_3 + S + 6AgF_2 \xrightarrow{\text{CCl}_4} NSF_3 + 3HF + 6AgF$$

Presumably the primary product in this reaction is NSF which, through further fluorination, goes over to NSF_3. In fact, one way of preparing NSF_3 is by passing NSF over AgF_2 at 100°C. (14):

$$NSF + 2AgF_2 \rightarrow NSF_3 + 2AgF$$

All reactions in which NSF is formed by fluorination [for example, from S_4N_4 (10)] also yield NSF_3 by further fluorination. The reaction of S_4N_4 with AgF_2 in boiling carbon tetrachloride to yield NSF_3 can be written as:

$$S_4N_4 + 12AgF_2 \xrightarrow{\text{CCl}_4} 4NSF_3 + 12AgF$$

In addition to NSF_3 a number of other compounds are formed, of which the following could either be isolated or identified from the infrared spectra: $N_4S_4F_4$, SiF_4, SF_4, SF_6, COF_2, SOF_2, SO_2F_2, CCl_3F, CCl_2F_2, SO_2, N_2, and Cl_2. These impurities must be carefully removed.

It has recently been reported (15) that S_2F_{10} reacts with a deficit of ammonia to give considerable amounts of NSF_3. Further information is lacking.

Procedure (11). 5 g. S_4N_4 and about 50–70 g. of freshly prepared silver difluoride are combined with 100 ml. carbon tetrachloride in the silica apparatus previously described in Sec. II.A.1. While being stirred, the mixture is rapidly heated to the boiling point of the carbon tetrachloride. The subsequent procedure is as described previously for the preparation of NSF. The reaction is complete after $1^1/_2$ hr. The crude product is washed with 50 ml. of a 3% $KMnO_4$ aqueous solution and purified by passing over finely divided lead dioxide. Further purification is accomplished by gas chromatography. Tricresylphosphate on kieselguhr is used as the stationary phase and hydrogen as the carrier gas.

Yield. Ca. 5 g. NSF_3 (corresponding to 24% of the theoretical value as calculated from the amount of S_4N_4 consumed).

2. Properties

NSF_3 is a colorless pungently smelling gas which condenses under normal pressure at -27.1 ± 0.1 °C. to a colorless liquid the freezing point of which is -72.6 ± 0.5 °C. The "chemical shift" of the fluorine nuclear magnetic resonance signal (using saturated aqueous potassium fluoride solution as a reference) lies at $+187$ p.p.m. (2). Characteristically strong infrared absorption bands lie at 1515, 811, 775, 521, 429, and 342 cm.$^{-1}$ and at frequencies corresponding to the combination bands (2).

In contrast to NSF, NSF_3 is very stable. In glass vessels at 200 °C. no reaction occurs. At 500 °C. a rapid reaction with the glass takes place with the formation of SiF_4, SO_2, N_2, sulfur, and metal fluorides. Gaseous hydrogen chloride does not react with NSF_3 upon heating; even gaseous ammonia is unreactive at room temperature and low pressures. Elementary sodium does not react with NSF_3 at room temperature either. By heating it to about 400 °C. it reacts to yield sodium sulfide, sodium fluoride, and nitrogen. In a electrodeless glow discharge, NSF_3 decomposes, yielding NSF, SF_4, and SF_6 (11).

Important for the purification of thiazyltrifluoride is the fact that it is stable against dilute acids. The gas reacts only slowly with water at room temperature. In aqueous solution it is decomposed only by heating it in dilute sodium hydroxide solution; during this decomposition the sodium salt of amidosulfuric acid is formed—which hydrolyzes in concentrated hydrochloric acid to ammonium ions and sulfate ions. These reactions can be used for the quantitative analysis of NSF_3.

III. CYCLIC COMPOUNDS

A. Tetrathiazyltetrafluoride, $N_4S_4F_4$

1. Preparation

Until now, tetrathiazyltetrafluoride, $N_4S_4F_4$, could be prepared only by fluorination of tetrasulfurtetranitride, S_4N_4 (1). Attempts to prepare $N_4S_4F_4$ in the same way as $N_3S_3F_3$ (e.g., by polymerization of NSF) have been unsuccessful. Consequently, it can be assumed that the fluorination of S_4N_4 to $N_4S_4F_4$ does not involve intermediate SN radicals but rather that the fluorine atoms attach themselves directly to the sulfur atoms in the S_4N_4 8-membered ring. By this reaction, the arrangement of the sulfur and nitrogen atoms of the 8-membered ring is considerably changed as compared with the starting material S_4N_4, and the aromatic character of the ring is lost, as shown in Figure 1.

The fluorination of S_4N_4 is accomplished with silver difluoride in carbon tetrachloride. It has been found that a copper reaction vessel and the use of silver difluoride containing CuF_2 favors the formation of the tetrafluoride (11). This reaction can be written as:

$$S_4N_4 + 4AgF_2 \xrightarrow{CCl_4} N_4S_4F_4 + 4AgF$$

Procedure (9,11). In the silica container described in Sect. II.A.1, 5 g. S_4N_4 and 50 g. of freshly prepared silver difluoride are mixed in 100 ml. carbon tetrachloride (which has been dried with calcium chloride and freshly distilled). The solution is slowly heated to the boiling point over a period of about 2 hr., as it is heated it turns green; then the mixture, while still hot, is rapidly filtered and the residue is extracted twice with hot carbon tetrachloride. Upon cooling overnight, the solution turns red. White needles of $N_4S_4F_4$, which should be filtered off quickly, precipitate. By evaporation of the filtrate, impure $N_4S_4F_4$ may be further obtained which can be purified by recrystallization in carbon tetrachloride.

Yield. 1 g. $N_4S_4F_4$ (14% of theoretical value as calculated from the amount of S_4N_4 consumed). By using copper containers higher yields may be obtained.

2. Properties

Tetrathiazyltetrafluoride, $N_4S_4F_4$, crystallizes as white needles which begin to decompose at 128°C. and melt at 153°C. The material can be sublimed undecomposed in high vacuum at 80°C. The solubility in carbon tetrachloride is 3.44 g. per liter at 2°C.; the solubility increases with temperature. The density at 20°C. is 2.326 g./cm³. Infrared bands (2) that can be used for identification of the substance lie at 1117, 786, 760, 709, 645, and 520 cm^{-1}. The "chemical shift" of the fluorine nuclear magnetic resonance signal (using saturated aqueous potassium fluoride solution as a reference substance) lies at +155 p.p.m.; only one absorption band is observed (2).

In moist air $N_4S_4F_4$ hydrolyzes slowly; in warm dilute sodium hydroxide solution it hydrolyzes rapidly following the equation:

$$N_4S_4F_4 + 8OH^- + 4H_2O \rightarrow 4NH_4F + 4SO_3{}^{2-}$$

In a sealed glass tube, $N_4S_4F_4$ decomposes at 100 °C. In addition to the main product, S_4N_4, the gaseous compounds SiF_4, SOF_2, and SO_2 are formed. By heating $N_4S_4F_4$ with AgF_2 in a copper container to 100 °C., NSF_3 forms in addition to other gaseous compounds.

B. Trithiazyltrifluoride, $N_3S_3F_3$

1. Preparation

If one allows NSF to stand for 3 days in a sealed glass container, a mixture of crystals is formed from which one can sublime the volatile $N_3S_3F_3$ (11). Thus, the expected polymerization of NSF to $N_3S_3F_3$ takes place, analogous to the polymerization of gaseous monomeric NSCl to $N_3S_3Cl_3$ (12).

$$3NSF \rightarrow N_3S_3F_3$$

The fluorination of $N_3S_3Cl_3$ with silver difluoride in carbon tetrachloride is a more convenient method of preparing $N_3S_3F_3$, since $N_3S_3Cl_3$ can readily be obtained from S_4N_4 and is easier to handle than NSF (16).

Procedure (15). In a small quartz vessel R (see Fig. 2) with a magnetic stirrer 20 g. AgF_2 is added to suspension of 10 g. of pulverized $N_3S_3Cl_3$ in 40 ml. of dried and freshly distilled carbon tetrachloride. A U-shaped tube filled with $CaCl_2$ prevents the admission of moisture. The reaction mixture has to be stirred at room temperature for about 18–20 hr. until the yellow solution becomes colorless. The reaction vessel is then attached to the filtration apparatus (Fig. 2) the joint of which has been lubricated with KEL-F and which has been flushed thoroughly with nitrogen. The colorless solution is separated from the precipitate of silver salts through a glass frit. The nitrogen stream is then turned off, and stopcocks H_1 and H_2 are closed. Subsequently, the collector A is attached to the apparatus (Fig. 3) and the filtrate is subjected to fractional condensation in high vacuum. For this purpose the collector trap A is cooled with a CO_2–acetone slurry to -80 °C. and evacuated briefly by opening stopcock H_1. After closing H_1 the traps B, C, D, and E are heated up in high vacuum. With continuous pumping, trap E is cooled to -180 °C. with liquid air, trap C to -40 °C. with a CO_2–methanol bath, and trap D is held at -20 °C. Stopcock H_1 is then opened again and, by lowering the CO_2–acetone bath, the substance in the trap slowly warms up and vaporizes. Strongly re-

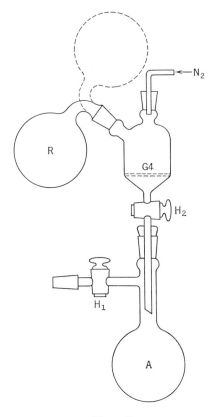

Figure 2

fracting colorless crystals of $N_3S_3F_3$ form in traps B and C; the carbon tetrachloride condenses in trap E. The pure $N_3S_3F_3$ is then sublimed from traps E and C to trap D. From trap D the substance can be further handled and condensed, for example, into analysis bulbs, melting point tubes for further investigation.

Yield. About 7 g., corresponding to 90% of the theoretical value as calculated from the amount of $N_3S_3Cl_3$ consumed.

2. *Properties*

$N_3S_3F_3$ forms colorless crystals which are markedly volatile even at room temperature. The melting point is 74.2°C., the boiling point 92.5°C. $N_3S_3F_3$ is readily soluble in carbon tetrachloride and benzene. The "chemical shift" of the fluorine nuclear magnetic resonance signal (using saturated aqueous potassium fluoride solution as a reference sub-

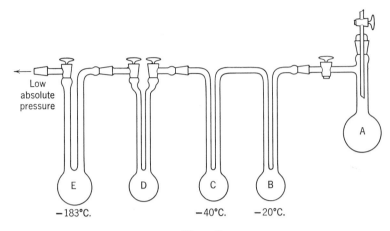

Figure 3

stance) lies at $+147$ p.p.m. (2); only one absorption band is observed. Infrared bands that can be used for the identification of the substance lie at 1085, 720, and 650 cm.$^{-1}$ (2). Although $N_3S_3F_3$ is stable in dry air, it decomposes and turns black in moist air. $N_3S_3F_3$ is much more sensitive toward moisture than $N_3S_3Cl_3$ and $N_4S_4F_4$. In cold dilute sodium hydroxide solution it hydrolyzes quantitatively according to the equation

$$N_3S_3F_3 + 6OH^- + 3H_2O \rightarrow 3NH_4F + 3SO_3{}^{2-}$$

C. Thiotrithiazylfluoride, S_4N_3F

1. Preparation (17)

The thiotrithiazyl compounds all contain the cationic 7-membered ring $S_4N_3{}^+$. The relatively stable chloride serves as the starting material for the preparation of these compounds, including the fluoride. To replace the chloride ion by the fluoride ion anhydrous hydrogen fluoride gas is reacted with S_4N_3Cl at 80–100°C. in a polyethylene or Teflon tube:

$$S_4N_3Cl + HF \rightarrow S_4N_3F + HCl$$

Yellowish-brown crystals of the composition $S_4N_3F \cdot 1.5$ HF are formed. The HF can be removed only by decomposing the compound.

Procedure. 0.5 g. S_4N_3Cl is placed in a polyethylene tube of about 30 cm. length provided with a Liebig condenser (18). The material is heated to 80–100°C. by pumping hot glycerine through the condenser. After the desired temperature has been reached, anhydrous hydrogen fluoride is passed over the chloride. A yellow liquid is formed and bubbles of HCl gas are evolved. When all the S_4N_3Cl has reacted, both the heating and the HF stream are turned off. The polyethylene tube is sealed off to a length of about 20 cm. near one end of the Liebig condenser. Under a protective CO_2 atmosphere the polyethylene tube is placed vertically in a vacuum system. After pumping off the excess HF, the sample is dried for another 3 hr. at 40°C. and then cooled. Finally, the apparatus is filled with dry CO_2.

2. Properties

S_4N_3F exists as brownish-yellow crystals which immediately become greenish-black in moist air. S_4N_3F is much more sensitive toward moisture than the chloride.

REFERENCES

1. Glemser, O., H. Schröder, and H. Haeseler, *Naturwissenschaften* **42**, 44 (1955); *Z. Anorg. Allgem. Chem.* **279**, 28 (1955).
2. Richert, H., and O. Glemser, *Z. Anorg. Allgem. Chem.* **307**, 328 (1961).
3. Kirchhoff, W. H., Dissertation, Harvard University, USA, 1962.
4. Kirchhoff, W. H., and E. B. Wilson, *J. Am. Chem. Soc.* **84**, 334 (1962).
5. Wiegers, G. A., and A. Vos, *Acta Cryst.* (*Copenhagen*) **14**, 562 (1961); **16**, 152 (1963).
6. Weiss, J., *Angew. Chem.* **74**, 216 (1962).
7. Craig, D. P., *Chem. Soc.* (*London*), *Spec. Publ.* **12**, 343 (1958); D. P. Craig, and N. L. Paddock, *Nature* **181**, 1052 (1958); D. P. Craig, *J. Chem. Soc.* **1959**, 997.
8. Glemser, O., E. Wyszomirski, and Horst Meyer, unpublished results.
9. Glemser, O., and S. Austin, unpublished results; S. Austin, thesis, Göttingen, 1961.
10. Glemser, O., and H. Richert, *Z. Anorg. Allgem. Chem.* **307**, 313 (1961).
11. Glemser, O., H. Meyer, and A. Haas, *Chem. Ber.*, in press. H. Meyer, Dissertation, Göttingen, 1964.
12. Glemser, O., and H. Perl, *Naturwissenschaften* **48**, 620 (1961).
13. Glemser, O., H. Schröder, and E. Wyszomirski, *Z. Anorg. Allgem. Chem.* **298**, 72 (1959).
14. Glemser, O., and H. Schröder, *Z. Anorg. Allgem. Chem.* **284**, 97 (1956); O. Glemser and H. Richert, *ibid.* **307**, 313 (1961).
15. Cohen, B., and A. G. MacDiarmid, *Chem. Ind.* (*London*) **1962**, 1866.
16. Schröder, H., and O. Glemser, *Z. Anorg. Allgem. Chem.* **298**, 78 (1959).
17. Glemser, O., and E. Wyszomirski, *Chem. Ber.* **94**, 1443 (1961).
18. For the preparation of S_4N_3Cl see A. Meuwsen, *Ber. Deut. Chem. Ges.* **65**, 1724 (1932).

CHAPTER 10

Hypohalites and Compounds Containing the −OX Group

STANLEY M. WILLIAMSON

University of California, Berkeley, California

CONTENTS

I. INTRODUCTION

This chapter will have as its main topic a presentation of the methods of preparation of that class of compounds known as hypohalites. The halogen of a hypohalite is "univalent" with its "valence" being satisfied by bonding to an oxygen atom. In some cases this X–O group will carry a negative charge as in ionic hypohalite salts (for example, Cl–O⁻ is the hypochlorite anion), and in others the second "valence" of the oxygen will participate in a covalent bond with another atom or group. The halogen compounds that are discussed under the title of "Halogen Cations" and that have recently been reviewed by Arotsky and Symons (1) will not be discussed here even though some would be positively univalent.

One of the first reactions with which a beginning chemistry student becomes acquainted is that of a halogen (not fluorine) and sodium hydroxide solution, from which the halide and hypohalite anions are obtained:

$$X_2 + 2OH^- \rightleftharpoons XO^- + X^- + H_2O \qquad\qquad 1$$

Perhaps he also learns that the action of fluorine on water or basic solution does not yield ionic species corresponding to those of the other halogens (2). Neither hypofluorous acid nor other ionic derivatives from the fluorine reaction have been observed, so the discussion of the reaction of fluorine to produce covalent hypofluorites will be given in a separate section.

From more detailed lectures and study, the student discovers that the oxidation potentials in aqueous solution for the various oxygen-containing halogen species do not favor the stability of the hypohalites or other intermediate oxidation states. Latimer (3) gives the following potential data for the chlorine system.

$$Cl^- + 2OH^- \rightarrow ClO^- + H_2O + 2e^- \qquad E^0 = -0.88 \qquad 2$$

$$ClO^- + 4OH^- \rightarrow ClO_3^- + 2H_2O + 4e^- \qquad E^0 = -0.50 \qquad 3$$

By appropriate combination of the two half reactions one gets

$$3ClO^- \rightarrow ClO_3^- + 2Cl^- \qquad E^0 = 0.38 \qquad 4$$

The positive standard potential for reaction **4** indicates that hypochlorite is thermodynamically unstable with respect to disproportionation. The corresponding equilibrium constant at room temperature, calculated from the relation $E^0 n\mathfrak{F} = RT \ln K$, is $10^{25.7}$. The same calculation for BrO^- gives $K = 10^{14.9}$ and for IO^-, $K = 10^{23.6}$, when appropriate potentials for bromine and iodine are used.

A more serious problem than the reaction of the hypohalite with itself is the reaction with halide if the pH of the solution is lowered. In acid solution, the halide and hypohalous acid will react to reverse the initial reaction and to produce water and free halogen.

The preparation of a pure hypohalite solution or salt must be accompanied by the removal of the by-product halide anion, e.g., by precipitation with silver ion. So it is highly desirable to carry out the halogenation of the base under such conditions that the impurities, e.g., halide and halate anions, are removed before the isolation of the hypohalite is attempted. Hypohalites appear, in general, to be very soluble, so low temperature evaporation of the solution is necessary for their isolation. If the evaporation is fast, then in some cases hypohalite salts have been obtained.

The decomposition of aqueous hypohalites has been studied. For HOCl (4a), BrO^- (4b), and IO^- (4c), the mechanism of decomposition appears to be second order in hypohalite. The rate determining step is $2XO^- \rightarrow XO_2^- + X^-$ which is followed by a rapid reaction of $XO_2^- + XO^- \rightarrow XO_3^- + X^-$. But in practice, this decomposition is not too rapid to prohibit the production and isolation of hypohalites except for hypoiodites.

II. IONIC HYPOHALITES

One technique used for the preparation of good quality LiOCl (5) involves the chlorination of a mixture of LiOH and NaOH solutions. The chloride was removed as NaCl before the more soluble LiOCl crystallized from the evaporating solution.

Another technique that appears to have quite general application has been used in the preparation of the hypobromite, $LiOBr \cdot 2H_2O$ (6). LiOH solution was treated with the stoichiometric amount of fresh, wet Ag_2O after a 1:1 molar ratio of LiOH to Br_2 was added. The regenerated LiOH was removed by titration with Dowex 50W-X8, 40-mesh H^+ form resin. The resulting LiOBr solution which was free from other ions and which had been kept cold and free of CO_2 was evaporated at reduced pressure. A bright-yellow solid with a purity of 95% could be obtained with an analysis of $LiOBr \cdot 2H_2O$. Hypobromite, bromite, and bromate anions can be determined accurately in the presence of each other (7), so analysis of the product is not difficult. At room temperature, the $LiOBr \cdot 2H_2O$ had only a very faint odor of bromine over the crystals and neither did the $[BrO_3^-]$ increase at an appreciable rate.

At the time of this writing, the isolation of crystalline hypoiodites does not seem to have been accomplished, and there are only three hypobromites, $LiOBr \cdot 2H_2O$ (6), $NaOBr \cdot 5H_2O$, and $KOBrO \cdot 3H_2$ (8). The odor of Br_2 over the latter two salts is appreciable at room temperature and the yellow orange color disappears rapidly.

An alternative method of preparing hypochlorites is the reaction of hypochlorous acid and a hydroxide. HOCl (9) which was prepared by reacting $Cl_2O_{(l)}$ with water can be obtained in concentrations upward to $5M$ and the relatively stable, white $Ca(OCl)_2$ has been obtained in purities of 94%.

Hypobromous acid solutions (6,10) of good purity have been made, but their lower stable concentrations tend to discourage their successful reaction with hydroxides to produce hypobromites in preparative amounts.

III. COVALENT HYPOHALITES

The first group of covalent hypohalites consists of the hypofluorites. Oxygen difluoride, OF_2, might be considered a difunctional hypofluorite and it is also the simplest with respect to structure. Dioxygen difluoride, O_2F_2, trioxygen difluoride, O_3F_2, and tetraoxygen difluoride, O_4F_2, are included even though their molecular structures have not been investigated in detail.

A number of the hypofluorites are produced by the replacement of the hydrogen atom in a strong acid by a fluorine atom. The members of this group are: fluorine nitrate, $FONO_2$; fluorine perchlorate, $FOClO_3$; fluorine fluorosulfonate, $FOSO_2F$; trifluoroacetyl hypofluorite (fluorine trifluoroacetate),

$$FOCCF_3$$
$$O$$

and pentafluoropropionyl hypofluorite (fluorine pentafluoropropionate),

$$FOCC_2F_5$$
$$O$$

The other members of the hypofluorite series are prepared by fluorination of certain oxides or oxyfluorides, the corresponding acids of which are unknown. These members are: trifluoromethyl hypofluorite, CF_3OF; pentafluorosulfur hypofluorite, SF_5OF; and pentafluoroselenium hypofluorite, SeF_5OF.

The hypofluorites are very reactive with respect to fluorination of other reagents. They are very volatile for their corresponding molecular weights and can be handled without difficulty (except for the explosive ones) in glass vacuum line systems. The compounds are not very sensitive to water which is indeed a fortunate property, so very careful exclusion of water vapor is not necessary in their preparation. Deliberate hydrolysis of a hypofluorite slowly yields oxygen, fluoride ion, and another anion, e.g., nitrate, perchlorate, fluorosulfonate, etc. The hypofluorites are strong enough oxidizing agents to react explosively with organic compounds, although in some cases the reaction has been moderated sufficiently to yield a new class of carbon compound derivatives (11–13).

IV. OXYGEN FLUORIDES

Oxygen difluoride, OF_2 (14–16), along with other compounds is the product of the direct fluorination of water. Fifty per cent conversions of F_2 to OF_2 have been accomplished by allowing the fluorine to bubble into $0.5M$ KOH at a depth of about 1 cm. The KOH, at as close to $0\,°C$. as possible, flows through the apparatus so that the concentration of OH^- will be a constant. Most of the OF_2 gas that is formed boils out of the solution and may be carried away by some excess unreacted fluorine and some formed oxygen. The solubility of OF_2 at 0°C. is 68 ml. (STP) per liter of water. The OF_2 gas is best trapped in glass U-traps open to the atmosphere at liquid oxygen temperature (-183°C.), since this temperature will condense neither F_2 nor air. If no carrier gas is used, the OF_2 will be fairly efficiently trapped at -183°C., even though it has an appreciable volatility at that temperature. It is very important that this reaction be carried out in a good fume-hood and in the absence of glass joints lubricated with hydrocarbon greases. The other oxygen fluorides have been prepared by means of an electric discharge taking place in the presence of different ratios of O_2 to F_2 at low pressures and temperatures. The basic procedures for the preparation of O_2F_2 (17), O_3F_2 (18), and O_4F_2 (19) are similar. A mixture of oxygen and fluorine is allowed to flow into the discharge apparatus at total pressures in the range of 5 to 15 mm. Hg. Products that are formed condense on the cold wall (ca. 90°K.) of the apparatus. It is interesting to note (19) that the preparation of O_4F_2 required lower discharge energies and temperatures than the preparations of O_2F_2 and O_3F_2. A drawing of the apparatus that has proved to be quite satisfactory can be found in Figure 1 of ref. (18a). It appears certain that this powerful preparative technique will find much wider application (20a). A current example can be found for its use in the preparation of KrF_4 (20b).

OF_2 is stable with respect to decomposition to about 250°C., but the higher oxyfluorides decompose in steps to the stable OF_2 at very low temperatures. This stepwise decomposition has been quite useful in the determination of the stoichiometry of the compounds.

V. HYPOFLUORITES OF STRONG ACIDS

The reaction of fluorine gas with strong acids is technically not complicated, and it has produced 5 hypofluorites. From an apparatus similar in design to that for OF_2, $FONO_2$ (21) has been obtained by fluorinating $3M$ HNO_3 at about 2°C. Because of the explosive nature of FNO_3, the apparatus was fabricated from nickel and Monel metals, and it was equipped with a blow-out opening as a safety precaution (21, see Fig. 2). A more convenient and efficient method appears to be the direct fluorination of $KNO_{3(s)}$ (22), which involves fluorine passing over KNO_3 contained in platinum or nickel boats. The excess F_2 carries the gaseous FNO_3 which solidifies in a trap at -183°C. and the F_2 passes on through. It is recommended that FNO_3 be made immediately before its planned use, because it has been observed to explode without apparent provocation. When 60% perchloric acid was fluorinated in a glass apparatus (23, see Fig. 1), the explosive gaseous product formed was identified as

FOClO$_3$. The fluorine flowed counter-current to the perchloric acid which was dripped over glass chips. A 90% yield of FOClO$_3$ was obtained when undiluted fluorine passed over 72% perchloric acid in a platinum boat. HF and FOClO$_3$ differ enough in volatility so that the FOClO$_3$ can be distilled from the mixture. The reaction of F$_2$ and HClO$_4$ appears to be rather complicated. When 60% HClO$_4$ was fluorinated in a carbon apparatus, no FOClO$_3$ was obtained, but, instead, about equal amounts of OF$_2$ and O$_2$ were obtained. The reaction of F$_2$ with KClO$_4$ as an alternative method of synthesis is not mentioned by the investigators.

Trifluoroacetyl hypofluorite, CF$_3$CO$_2$F (24), and pentafluoropropionyl hypofluorite, C$_2$F$_5$CO$_2$F (25), have been prepared from the vapor-phase fluorination of trifluoroacetic and pentafluoropropionic acids, respectively. The apparently necessary presence of water seems to indicate that the reaction mechanism is quite complicated. The dependence on the rates of formation on water has not been elucidated at this time. These two hypofluorites are quite unstable and can be exploded quite violently if subjected to a spark. The decomposition products are the saturated fluorocarbon and carbon dioxide. This fluorination reaction does point out that room temperature fluorinations do not always require elaborate and nonreactive apparatus. The rubber, polyethylene, and Saran parts of the apparatus (24, Fig. 1) did not appear to produce an objectionable amount of fluorocarbons.

VI. HYPOFLUORITES OF OXIDES

Three hypofluorites have been prepared by means of "catalytic" fluorination (26). They are trifluoromethyl hypofluorite, CF$_3$OF (27), fluorine fluorosulfonate, FOSO$_2$F (28), and pentafluorosulfur hypofluorite, F$_5$SOF (29). Either methanol, carbon monoxide, or carbonyl fluoride can be the starting material for the production of CF$_3$OF. At 170°C. methanol and carbon monoxide gave CF$_3$OF in 50 and 70% yields, respectively. Unless excess fluorine is used, e.g., F$_2$/CO = 5, some COF$_2$ is formed. The ready availability of CO makes the fluorination of COF$_2$ unwise. Also, the use of CO eliminates the necessity of removing HF, the byproduct from CH$_3$OH. It appears that COF$_2$ and CF$_3$OF cannot be separated by distillation because of azeotrope formation. The contaminant COF$_2$ can easily be removed from the CF$_3$OF by reaction with water. At room temperature, COF$_2$ rapidly gives CO$_2$ and HF by hydrolysis; whereas, CF$_3$OF oxidizes water only slowly.

The "catalytic" reactor is essentially only a copper pipe surrounded by heating wire and insulation with appropriately placed thermocouples inside or outside the copper pipe. The size of the reactor is a function of the desired flow rate and residence time for the gases. A typical size reactor could be a meter in length with an unpacked volume of 1 to 4 l. The packing can be copper ribbon, turnings, or even screen wire which has been silver plated. Kitchen variety copper "chore balls" are quite satisfactory because they pack evenly. Quarter-inch copper tubing with silver solder connections to the main reactor pipe serve as inlets and outlets. The reactor can be operated at reduced, atmospheric, or elevated pressures, if suitable auxiliary equipment is connected.

The catalytic effect of AgF_2 is not well understood in the formation of hypofluorites. AgF_2 by itself is not capable of fluorinating CO or the other oxides to the hypofluorite. Elemental fluorine is necessary in addition to the AgF_2.

At 220°C., SO_3 is fluorinated in the presence of AgF_2 and excess fluorine to $FOSO_2F$ in yields of 60%. The remainder of the sulfur is converted to sulfuryl fluoride. A temperature of 100°C. and limited fluorine with SO_3 produce peroxydisulfuryl difluoride, $S_2O_6F_2$ (30). SO_3F_2 and $S_2O_6F_2$ can easily be separated by distillation. Their boiling points are -31.3 and $+64.1$°C., respectively. SO_2F_2 which boils at -56.2°C. is also easily separable. The reaction of F_2 with K_2SO_3 as an alternative method of synthesis is not mentioned by the investigators.

At 215°C., thionyl fluoride, SOF_2, is readily fluorinated "catalytically" to pentafluorosulfur hypofluorite, F_5SOF. Sulfur hexafluoride and sulfuryl fluoride are the major by-products, and the hypofluorite can be obtained in yields upward from 50% based on the SOF_2 consumed. With a boiling point of -35.1°C., the hypofluorite can readily be purified by distillation from the more volatile contaminants.

It is significant to mention here that $SOCl_2$ is an unsatisfactory reagent for fluorination to SF_5OF. It must first be converted to SOF_2 (31); if not, the chlorine will be absorbed by the silver in the reactor and, also, some Cl_2 will appear mixed with the F_5SOF, making separation very difficult.

The last hypofluorite in this discussion has been obtained in small yields from the direct fluorination, at 60 to 90°C., of selenium dioxide or selenium oxychloride. It is pentafluoroselenium hypofluorite, F_5SeOF (32).

VII. HALOGEN(I) OXIDES

As has been pointed out earlier, the action of chlorine and bromine on aqueous basic materials produces the corresponding hypohalites; whereas, for fluorine, the hypofluorous acid anhydride, F_2O, is formed. The corresponding Cl_2O (9) and Br_2O (33) species can be obtained by reaction of the appropriate halogen with freshly prepared yellow mercuric oxide. Cl_2O is stable enough to be stored as a reddish-brown liquid somewhere below its boiling point of $+2°C$. Cl_2O is very soluble in CCl_4 and can be stored as a solution therein. The brown Br_2O is best stored in CCl_4 solution in the region of $-30°C$. The independent existence of I_2O appears to be quite doubtful. There is a very nice example of chemical periodicity in the halogen monoxides, when one considers the relative stability of the species.

VIII. HALOGEN(I) NITRATES

The preparation of fluorine nitrate (21) has already been described. There are also nitrates of the other halogens to complete the series. They are: chlorine(I) nitrate, $ClONO_2$ (34), bromine(I) nitrate, $BrONO_2$ (35), and iodine(I) nitrate, $IONO_2$ (35). There is no evidence to doubt that the molecular structures of $BrONO_2$ and $IONO_2$ are different from each other or different from the molecular structures of $FONO_2$ and $ClONO_2$ (36), which are like that of nitric acid.

$ClONO_2$ is the product of almost any reaction of Cl_2O with a nitrogen oxide. The cases studied have been the reaction of Cl_2O and N_2O_5 or N_2O_4. The N_2O_4 reaction gives a mixture of products, $ClONO_2$ and $ClNO_2$. The $ClONO_2$ is a pale-yellow liquid which boils at $18°C$. The "positive" character of the chlorine has been demonstrated by reaction with NaCl to give Cl_2 and $NaNO_3$. The reaction of $ClONO_2$ with unsaturated hydrocarbons at low temperatures has been investigated (37), wherein the Cl and NO_3 groups add directly to the double bond to give 2-chloroalkyl nitrates. The reaction is analogous to that observed for some hypofluorites (11–13). Reaction of BrCl with $ClONO_2$ gives $BrONO_2$ and Cl_2 at $-70°C$. The yellow $BrONO_2$ melts at $-42°C$. and begins to decompose at about $0°C$. into Br_2, NO_2, and O_2. Reactions of chlorine nitrate with ICl_3 produce $I(NO_3)_3$ and addition of excess I_2 in $CFCl_3$ solvent below $0°C$. gives a yellow solid, $IONO_2$, which decomposes at about $0°C$. The bromine(III) and iodine(III) nitrates are clearly more stable than the bromine(I) and iodine(I) species.

IX. HALOGEN(I) FLUOROSULFONATES

The production of fluorine fluorosulfonate by direct synthesis (28) from F_2 and SO_3 has been discussed. An alternative method of preparation (38) in which $S_2O_6F_2$ (30) is fluorinated at $250°C$. has also proved successful with different conditions in the preparation of the entire series of the halogen(I) fluorosulfonates.

In the original work (38), it was found that if excess Br_2 or I_2 was re-acted with liquid $S_2O_6F_2$ at room temperature in glassware, $BrOSO_2F$ and probably $IOSO_2F$ were formed. Excess $S_2O_6F_2$ gave Br(III) and I(III) fluorosulfonates quantitatively, to which, the addition of extra halogen gave the halogen(I) species. $BrOSO_2F$ is a reddish-black viscous liquid with a normal boiling point of 120.5°C. The compound is stable with respect to decomposition, but reacts quite vigorously with water. $I(SO_3F)_3$ is a pale-yellow liquid with a freezing point at 32.2°C. Heating the compound to 114°C. at 30 mm. Hg produces $S_2O_6F_2$ and a non-volatile green liquid. The same green-colored liquid can be observed when a limited amount of $S_2O_6F_2$ is reacted at room temperature with I_2. The I(I) species is apparently very unstable at room temperature, if even formed at all by the reaction in which variable stoichiometry is observed.

The most difficult halogen fluorosulfonate to prepare has been the chlorine derivative, $ClOSO_2F$ (11). Colorless $S_2O_6F_{2(l)}$ vaporizes and the vapor at 125°C. is quite dark brown. This color is thought to be due to the SO_3F radical. At moderate pressure, SO_3F does not react with Cl_2; but when Cl_2 and $S_2O_6F_2$ were distilled into a Monel metal tube fitted with a valve in a molar ratio of 1.5:1 and heated in the liquid state at 125°C. for 5 days, nearly pure $ClOSO_2F$ was produced. The very reactive yellow liquid product darkens toward red upon standing in glass at room temperature.

X. HALOGEN(I) PERCHLORATES

Recent work in nonaqueous solvents (39) has shown that, at −85°C. in ethanol, iodine and silver perchlorate react to give AgI and iodine(I) perchlorate, $IOClO_3$. Earlier work, mostly at room temperature, in ether was complicated by iodination of the solvent and from it the for-mation of ClO_4 (40) was incorrectly postulated. The reaction of bromine and silver perchlorate in aromatic solvents has been studied (40) and bromination of the solvent was observed. The work did not state that $BrOClO_3$ might have been the active reagent. Since $IOClO_3$ was not isolated from the alcohol solvent, $FOClO_3$ (23) has the distinction of being the only halogen(I) perchlorate to have been isolated.

XI. HALOGEN(I) ALKOXIDES

The first literature on organic hypohalites appeared in 1885 and was concerned with the preparations of methyl- and ethylhypochlorites (41) by direct chlorination of the alcohols in the presence of sodium hydroxide solution. A general preparative procedure (42) for the oily, yellowish hypochlorites involves the chlorination of a cool solution of 2 moles of

NaOH per mole of alcohol. The water-insoluble ROCl layer can be separated, washed, and distilled.

The primary and secondary alkyl hypochlorites are very unstable (loss of HCl to give a carbonyl compound). In the pure state they are so unstable that the exposure to light will ignite them, sometimes explosively. A cold, dark environment seems to be the only way to prepare the pure material.

Tertiary butyl hypochlorite, as are the other tertiary alkyl hypochlorites, is remarkably stable. The yellow liquid can be stored at ordinary temperatures for months without decomposition if bright light is excluded. It can be distilled at normal pressure without change (79.6 °C. at 750 mm. Hg).

An alternative method of preparation (43) involves the mixing together and shaking of a solution of hypochlorous acid and an alcohol dissolved in an unreactive, water-insoluble solvent, e.g., CCl_4. The primary and secondary hypochlorites are more stable in solution than in the pure form. Alcohols will also react with a CCl_4 solution of Cl_2O to give solutions of hypochlorites.

In a kinetic study on the formation of t-BuOCl (44), it was observed that OCl^- was not the species that produced the ROCl compound. Only at a pH less than 10, where the HOCl concentration is significant, is there appreciable yield of ROCl. It had been observed earlier that the chlorination of a basic solution of an alcohol produced no hypochlorite until most of the base had been destroyed by chlorine.

A good review (45) is available in which the organic hypohalites, preparations, chemistry, and other references are given in detail. A recent word of caution on the explosive nature of t-BuOCl has appeared (46).

The preparation of pure t-butyl hypobromite (10) has been accomplished by reacting aqueous HOBr (Br^- free) with a solution of t-BuOH in trichlorofluoromethane in a 43% yield. Direct bromination of a basic solution of t-BuOH did not yield any t-BuOBr.

The reddish-orange liquid t-BuOBr was isolated from the Freon-11 solvent by vacuum distillation, b.p.: 44–45 °C./85 mm. The hypobromite is stable for long periods in the cold and dark and decomposes either in strong light, above 85 °C., or in the presence of a base.

There does not appear to be a report on the preparation and isolation of an organic hypoiodite.

XII. HALOGEN(ı) CARBOXYLATES

The acyl hypohalites with the exception of the above-mentioned perfluoroacyl hypofluorites (24,25) have been studied and written about to a large extent. Most of the work mentions the acyl hypohalites as intermediates in the reaction of a halogen with a metal salt of a carboxylic acid. The halogenation of the silver salt of carboxylic acids, the so-called Hunsdiecker reaction, has been the subject of several recent, good reviews (47).

In this reaction, under anhydrous conditions in an unreactive organic solvent such as CCl_4 or $C_6H_5NO_2$, when equimolecular amounts of X_2 and RCO_2Ag are mixed, the intermediate acyl hypohalite decarboxylates to give the corresponding alkyl halide, carbon dioxide, and the initially formed silver halide. Dilute CCl_4 solutions of acetyl hypochlorite and hypobromite have been studied spectrophotometrically (48), wherein the presence of the O–X bonding linkage was firmly established. These solutions were prepared by shaking a Cl_2/CCl_4 solution with mercuric or silver acetate or by adding excess acetic acid to the corresponding halogen monoxide.

Silver salts of perfluorocarboxylic acids with halogens yield more stable hypohalites than the normal acyl derivatives. This is supported by the observation that decarboxylation of CF_3CO_2Ag and I_2 does not occur appreciably below 100°C. (49); whereas, the hydrogen carboxylic acid salts lose CO_2 quite often at room temperature. With the hypohalites of the perfluorocarboxylic acids and some hypofluorites, there are representatives for all of the halogens, but in the less stable hypohalites of the hydrogen carboxylic acids, the hypofluorite member is absent.

The author has knowingly omitted many references which are concerned with phases of hypohalite chemistry other than general background and synthesis. It is his hope that the references given will provide the practicing inorganic chemist with sufficient information so that he can use the hypohalites in even larger areas of research.

In general, if a more thorough understanding is desired, the literature cited in the references to this chapter contain data on related topics other than background and synthesis.

REFERENCES

1. Arotsky, J., and M. C. R. Symons, *Quart. Revs.* **16**, 282 (1962).
2. Cady, G. H., *J. Am. Chem. Soc.* **56**, 1647 (1934).
3. Latimer, W. M., *The Oxidation States of the Elements and Their Potentials in Aqueous Solutions*, 2nd ed., Prentice-Hall, Englewood Cliffs, New Jersey, 1952, pp. 53–69.
4a. Lister, M. W., *Can. J. Chem.* **30**, 879 (1952); b. P. Engel, A. Oplatka, and B. Perlmutter-Hayman, *J. Am. Chem. Soc.* **76**, 2010 (1954); c. C. H. Li and C. F. White, *J. Am. Chem. Soc.* **65**, 335 (1943).
5. Cady, G. H., U. S. Patent 2,356,820 (to Pittsburgh Plate Glass Co.); through *Chem. Abstr.* **39**, 163 (1945).

250   S. M. WILLIAMSON

6. Williamson, S. M., "An Investigation into the Preparation of Crystalline Hypo-bromites," Technical Report, Physical Research Laboratory, The Dow Chemical Company, Midland, Michigan, 1960.
7. Chapin, R. M., *J. Am. Chem. Soc.* **56**, 2211 (1934).
8a. Scholder, V. R., and K. Krauss, *Z. Anorg. Allgem. Chem.* **268**, 279 (1952); b. also see, *Nouveau Traite de Chimie Minerale*, Vol. XVI, P. Pascal, ed., Masson et Cie, Paris, 1960, p. 430.
9. Cady, G. H., in *Inorganic Syntheses*, Vol. 5, T. Moeller, ed., McGraw-Hill, New York, 1957, p. 156.
10. Walling, C., and A. Padwa, *J. Org. Chem.* **27**, 2976 (1962).
11. Gilbreath, W. P., and G. H. Cady, *Inorg. Chem.* **2**, 496 (1963).
12. Williamson, S. M., and G. H. Cady, *Inorg. Chem.* **1**, 673 (1962).
13. Allison, J. A. C., and G. H. Cady, *J. Am. Chem. Soc.* **81**, 1089 (1959).
14. Lebeau, P., and A. Damiens, *Compt. Rend.* **185**, 652 (1927).
15. Cady, G. H., *J. Am. Chem. Soc.* **57**, 246 (1934).
16. Yost, D. M., in *Inorganic Syntheses*, Vol. 1, H. J. Booth, ed., McGraw-Hill, New York, 1939, p. 109.
17. Ruff, O., and W. Menzel, *Z. Anorg. Allgem. Chem.* **211**, 204 (1933).
18a. Kirshenbaum, A. D., and A. V. Grosse, *J. Am. Chem. Soc.* **81**, 1277 (1959); b. also see, S. Aoyama and S. Sakuraba, *J. Chem. Soc.* (*Japan*) **59**, 1321 (1938); c. *ibid.* **62**, 208 (1941).
19. Grosse, A. V., A. G. Streng, and A. D. Kirshenbaum, *J. Am. Chem. Soc.* **83**, 1004 (1961).
20a. Massey, A. G., *J. Chem. Ed.* **40**, 311 (1963); b. A. V. Grosse, A. D. Kirshen-baum, A. G. Streng, and L. V. Streng, *Science* **139**, 1047 (1963).
21. Cady, G. H., *J. Am. Chem. Soc.* **56**, 2635 (1934).
22. Yost, D. M., and A. Beerbower, *J. Am. Chem. Soc.* **57**, 782 (1935).
23. Rohrback, G. H., and G. H. Cady, *J. Am. Chem. Soc.* **69**, 677 (1947).
24. Cady, G. H., and K. B. Kellogg, *J. Am. Chem. Soc.* **75**, 2501 (1953).
25. Menefee, A., and G. H. Cady, *J. Am. Chem. Soc.* **76**, 2020 (1954).
26. Cady, G. H., A. V. Grosse, E. J. Barber, L. L. Burger, and Z. D. Sheldon, *Ind. Eng. Chem.* **39**, 290 (1947).
27. Kellogg, K. B., and G. H. Cady, *J. Am. Chem. Soc.* **70**, 3986 (1948).
28. Dudley, F. B., G. H. Cady, and D. F. Eggers, Jr., *J. Am. Chem. Soc.* **78**, 290 (1956).
29. Dudley, F. B., G. H. Cady, and D. F. Eggers, Jr., *J. Am. Chem. Soc.* **78**, 1553 (1956).
30. Dudley, F. B., and G. H. Cady, *J. Am. Chem. Soc.* **79**, 513 (1957).
31a. Booth, H. S., and F. C. Mericola, *J. Am. Chem. Soc.* **62**, 640 (1940); b. C. W. Tullock and D. D. Coffman, *J. Org. Chem.* **25**, 2016 (1960).
32. Mitra, G., and G. H. Cady, *J. Am. Chem. Soc.* **81**, 2646 (1959).
33. *Nouveau Traite de Chimie Mineral.* Vol. XVI, P. Pascal, ed., Masson et Cie, Paris, 1960, p. 423.
34a. Martin, H., and E. Kohnlein, *Z. Physik. Chem.* (*Frankfurt*) **17**, 375 (1958); b. M. Schmeisser and K. Brandle, *Angew. Chem.* **73**, 388 (1961); c. W. Fink, *Angew. Chem.* **73**, 466 (1961).
35. Schmeisser, M., and L. Taglinger, *Chem. Ber.* **94**, 1533 (1961).
36a. Skiens, W. E., and G. H. Cady, *J. Am. Chem. Soc.* **80**, 5640 (1958); b. K. Brandle, M. Schmeisser, and W. Luttke, *Chem. Ber.* **93**, 2300 (1960).
37a. Fink, W., *Angew. Chem.* **73**, 532 (1961); b. also see, ref. 32c.

38. Roberts, J. E., and G. H. Cady, *J. Am. Chem. Soc.* **81,** 4166 (1959).
39. Alcock, N. W., and T. C. Waddington, *J. Chem. Soc.* **1962,** 2510.
40. Gomberg, M., *J. Am. Chem. Soc.* **45,** 398 (1923).
41. Sandmeyer, T., *Chem. Ber.* **18,** 1767 (1885).
42a. Chattaway, F. D., and O. G. Backeberg, *J. Chem. Soc.* **1923,** 2999; b. H. H. Teeter and E. W. Bell, in *Organic Syntheses*, Vol. 32, R. T. Arnold, ed., Wiley, New York, 1952, p. 20.
43. Taylor, M. C., R. B. MacMullin, and C. A. Gammal, *J. Am. Chem. Soc.* **47,** 395 (1925).
44. Anbar, M., and I. Dostrovsky, *J. Chem. Soc.* **1954,** 1094.
45. Anbar, M., and D. Ginsburg, *Chem. Rev.* **54,** 925 (1954).
46. Bradshaw, C. P. C., and A. Nechuatal, *Proc. Chem. Soc.* **1963,** 213.
47a. Kleinberg, J., *Chem. Rev.* **40,** 381 (1947); b. R. G. Johnson and R. K. Ingham, *Chem. Rev.* **56,** 219 (1956); c. C. V. Wilson, in *Organic Reactions*, Vol. IX, R. Adams, ed., Wiley, New York, 1957, p. 332.
48. Anbar, M., and I. Dostrovsky, *J. Chem. Soc.* **1954,** 1105.
49. Henne, A. L., and W. F. Zimmer, *J. Am. Chem. Soc.* **73,** 1362 (1951).

AUTHOR INDEX*

A

Abedini, M., 170 (ref. 39), 171 (refs. 50, 53), 176 (ref. 76), 181 (refs. 50, 53), 183 (ref. 53), 192 (ref. 76), 193 (ref. 39), 194, *199, 200*

Abel, E. W., 79 (ref. 5a), *116*

Adams, W. B., Jr., 97 (ref. 95), *119*

Addison, C. C., 96 (ref. 81), *118*, 141, 142 (refs. 1, 3, 5, 9, 10, 13–15), 143 (refs. 13, 15, 23, 24), 144 (ref. 3), 145 (refs. 3, 5, 14, 29–32), 146 (refs. 31, 33), 147 (refs. 32, 34, 35), 149 (refs. 36, 39), 150 (refs. 40–42, 44, 45), 152 (ref. 35), 153 (refs. 5, 10, 32, 36, 46–49, 51), 154 (refs. 10, 48), 155 (ref. 14), 158 (ref. 1), 159 (refs. 55, 56), 160 (refs. 55, 56), *162–164*

Aftandilian, V. D., 144 (ref. 28), *163*

Alam, A., 15 (ref. 30), 16, *26*

Alberola, A., 78, 81 (ref. 19), 97 (refs. 19, 102), 107 (refs. 19, 102), 114 (ref. 19), *116, 119*

Albert, C., 172 (ref. 55), *200*

Albert, P., 212, 213, *224*

Albinak, M. J., 31–34 (ref. 8), *52*

Alcock, N. W., 247 (ref. 39), *251*

Alire, R. M., 63 (ref. 15), *74*

Allegra, G., 78, 81 (ref. 19), 97 (ref. 19), 107 (ref. 19), 114 (ref. 19), *116*

Allen, E. A., 127 (ref. 36), 128 (refs. 36, 40), 137 (ref. 36), 138 (ref. 36), *140*

Allen, J., 143 (ref. 23), *163*

Allison, J. A. C., 242 (ref. 13), 246 (ref. 13), *250*

Allpress, J. G., 142 (ref. 7), *162*

Amma, E. L., 31 (ref. 9), *52*

Amon, W. F., Jr., 7, *25*

Anacreon, R. E., 116 (ref. 140), *120*

Anbar, M., 248 (refs. 44, 45), 249 (ref. 48), *251*

Andersen, S., 51 (ref. 133), *56*

Anderson, H. H., 166 (ref. 13), 170 (refs. 40–42), 171 (ref. 45), 173 (ref. 42), 174 (ref. 42), 175 (ref. 70), 178 (ref. 89), 186 (ref. 110), 187 (refs. 110–112), 188 (ref. 89), 189 (refs. 110, 122), 190 (refs. 123, 124, 126, 127), 191 (refs. 133, 142), 196 (ref. 13), *198–202*

Anderson, J. S., 79 (ref. 2), 106 (ref. 2), *116*

Angell, C., 112 (ref. 117), 113 (ref. 117), *119*

Anisimov, K. N., 78, 83 (ref. 22), 96 (refs. 54, 66, 70, 83, 85), 97 (refs. 54, 83, 106), *116–119*

Antipin, P. F., 197 (ref. 151), *202*

Antsyshkima, A. S., 69 (ref. 43), *75*

Aoyama, S., 243 (ref. 18), *250*

Appenrodt, J., 218 (ref. 36), *224*

Arago, E., 31, *52*

Archer, R., 50 (ref. 103), *55*, 63 (ref. 18), 71, *74*

Arlitt, H., 42 (ref. 48c), *54*

Arnstein, H. R. V., 43 (ref. 53), *54*

Arotsky, J., 239, *249*

Arthur, P., Jr., 96 (ref. 42), *117*

Asato, G., 112 (ref. 117), 113 (ref. 117), *119*

Audrieth, L. F., 143 (ref. 22), *163*

Austin, S., 229 (ref. 9), 233 (ref. 9), *237*

Auten, R. W., 46, 49 (ref. 80), *55*

Aylett, B. J., 176 (ref. 84), 183 (ref. 84), 184, *200*

B

Babchinitser, T. M., 15 (ref. 32), 16 (ref. 32), *26*

Backeberg, O. G., 247 (ref. 42), *251*

Badische Anilin- und Soda-Fabrik A.G., 96 (refs. 63, 75), *118*

Bailar, J. C., Jr., 1, 6 (ref. 17), 11, 12, 14, 15 (ref. 24), 21, 23, *25–27*, 39 (ref. 36), 40, 41, 42 (refs. 46a, 46b, 46d, 46e), 44 (ref. 67), 46, 49 (refs. 40, 80), 50 (refs. 40, 91, 95, 96, 103, 107, 111, 112), 51 (refs. 36, 91), *53–56*, 62 (refs. 6, 8), 68 (ref. 38), 69 (ref. 42), 73 (ref. 42), 74 (ref. 42), *74, 75*

* *Italic* numbers refer to reference pages.

K

Kaczmarczyk, A., 189 (ref. 121), 197 (ref. 146), *201, 202*

Kaesz, H. D., 78, 81 (ref. 21), 94 (refs. 38 40a), 97 (refs. 20, 21, 38), 102 (ref. 20a), 103 (ref. 38), 110 (ref. 38), 112 (refs. 20, 38), 113 (refs. 20, 21, 38), 114 (ref. 38), *116, 117, 119*

Kahlen, N., 188 (ref. 119), *201*

Kane, M. W., 7, *25*

Karagounis, G., 42, 43, *54*

Kasarnowsky, J., 208 (ref. 4), *223*

Katz, J. J., 149 (ref. 37), 156 (ref. 53), 157 (refs. 37, 53), *163*

Katzin, L. I., 158 (ref. 52), *163*

Kauzmann, W., 32 (ref. 17), 33 (ref. 17), *53*

Kawai, K., 97 (ref. 88), *118*

Kawana, Y., 221, *224*

Kedzia, B., 149 (ref. 38), *163*

Keeley, D. F., 96 (ref. 47), 98, 99 (ref. 47), 100, *117*

Keggin, J. F., 2 (ref. 3), *25*

Kellogg, K. B., 244 (refs. 24, 27), 248 (ref. 24), *250*

Kennard, S. M. S., 186 (ref. 104), *201*

Kennedy, C. D., 128 (ref. 43), 138 (ref. 43), 139 (ref. 43), *140*

Kennelley, J. A., 222 (ref. 58), 223 (ref. 58), *225*

Kewley, R., 190 (refs. 130, 131), *202*

Keyes, F. G., 216, 219 (ref. 31), *224*

Kikkawa, I., 43 (ref. 56), *54*

Kilner, M., 96 (ref. 81), *118*, 142 (refs. 3, 5), 144 (ref. 3), 145 (refs. 3, 5), 153 (ref. 5), *162*

Kilpatrick, M., 3 (ref. 5), *25*

King, R. B., 91, 96 (ref. 71c), 99, *117, 118*

Kirch, L., 96 (ref. 45), *117*

Kirchhoff, W. H., 228 (refs. 3, 4), *237*

Kirkwood, J., 32 (ref. 17), 33 (ref. 17), *53*

Kirschner, S., 29, 31, 32–34 (ref. 8), 38 (ref. 11), 40 (refs. 3, 4), 41, 43, 46, 49 (ref. 40), 50 (refs. 40, 62), 51 (refs. 3, 4), *52–54, 57*

Kirshenbaum, A. D., 243 (refs. 18–20), *250*

Kittleman, E. T., 62 (ref. 10), 64 (ref. 10), 68 (ref. 10), 73 (refs. 10, 46), *74, 75*

Klein, C. H., 166 (ref. 12), 195 (ref. 12), *198*

Klein, R., 11 (ref. 24), 14, 15 (ref. 24), *26*

Kleinberg, J., 249 (ref. 47), *251*

Klemm, L. H., 43, *54*

Klemm, W., 128 (ref. 38), *140*

Kluiber, R. W., 11, 12, *26*

Klyne, W., *57*

Knopf, E., 41 (ref. 43), 45 (ref. 43), *53*

Knoth, W. H., 175 (ref. 73), 188 (ref. 73), *200*

Kobayashi, G., 94, *117*

Kobayashi, M., 41 (refs. 44b, 45), 45 (ref. 44b), *53*

Kohnlein, E., 246 (ref. 34), *250*

Kolobova, N. E., 78, 83 (ref. 22), *116*

Kolomnikov, I. S., 78, 83 (ref. 22), *116*

Korshak, V. V., 15 (ref. 32), 16, *26*

Korst, W. L., 222 (ref. 53), *225*

Koster, G. L., 48 (ref. 83), 50 (ref. 83), *55*

Kotake, M., 43 (ref. 57), *54*

Krajewski, A., 212, *223*

Krauss, K., 241 (ref. 8), *250*

Krebs, H., 42, 43, 50 (ref. 48a), *54*

Krill, H., 176 (ref. 79), *200*

Kriner, W. A., 174 (ref. 66), 179 (ref. 66), *200*

Kuchen, W., 189 (ref. 120), *201*

Kuebler, J. R., Jr., 42 (ref. 46d), *54*

Kuhn, W., 32 (ref. 17), 41, 45 (ref. 43), *53*

Kuljian, E. S., 175 (ref. 74), *200*

Kummer, D., 170 (ref. 37), 176 (refs. 37, 80–83), 182 (refs. 80, 82), 183, 184 (ref. 82), *199*

Kupa, G., 123 (ref. 15), *139*

Kuroya, H., 41 (ref. 45), *53*

L

Ladell, J., 114 (ref. 134), *120*

Lafferty, J. M., 167 (ref. 22), *199*

Lagally, H., 78, 96 (refs. 14, 71a), 97 (ref. 16), *116, 118*

Landa, S., 213, 217, 219 (ref. 24), *224*

Langer, A., 17 (ref. 33), *26*

Langer, C., 78, 97 (ref. 6), *116*

Latimer, W. M., 240, *249*

Laudise, R. A., 128 (ref. 44), 136 (ref. 51), *140*

Lavrent'ev, V. N., 83 (ref. 30a), *117*

Lawton, E. A., 24 (ref. 43), *27*

Nicholls, D., 122 (ref. 2), 123 (ref. 9), 136 (ref. 2), 137 (ref. 2), *139*
Nielsen, N. C., 51 (ref. 119), *56*
Nixon, E. R., 188 (ref. 116), *201*
Noack, K., 111–113 (ref. 111), *119*
Norbury, A. H., 159 (ref. 55), 160 (ref. 55), *164*
Nunez, L. J., 64, *74*
Nurok, D., 43 (ref. 61), *54*
Nuss, J. W., 197 (ref. 145), *202*
Nyholm, R. S., 83, 111 (refs. 109, 112, 115), 112 (refs. 112, 115), 113 (ref. 115), *117*, *119*, 125 (ref. 26), *139*

O

O'Brien, R. J., 142 (ref. 4), *162*
O'Brien, T. D., 50 (ref. 111), *56*
Ocone, L. R., 17 (ref. 34), *26*, 93, *117*
O'Connell, J. J., 61 (ref. 2), 63 (ref. 2), *74*
O'Dwyer, M. F., 38 (ref. 25), 48 (ref. 25), *53*
Oehmke, R. W., 69 (ref. 42), 73 (ref. 42), 74 (ref. 42), *75*
Offermann, W., 61 (ref. 5), *74*
Oh, J. S., 11, 12, 15 (ref. 24), *26*
Okhlobystin, O. Yu., 96 (refs. 58a, 58b), 97 (refs. 58a, 58b), *117*, *118*
Olszewski, E. J., 69 (ref. 42), 73 (ref. 42), 74, *75*
Onyszchuk, M., 173 (refs. 61, 63), 174 (ref. 61), 178 (ref. 63), 179, 189 (ref. 120), *200*, *201*
Opgenhoff, P., 127 (ref. 32), *139*
Opitz, H. E., 182 (ref. 93), *201*
Oplatka, A., 240 (ref. 4b), *249*
Orchin, M., 96 (refs. 40b, 45), 100 (ref. 40b), *117*
Orlov, N. F., 174 (ref. 67), *200*
Ormont, B., 82, *117*
Osborne, B. P., 128 (ref. 48), *140*
Owen, B. B., 96 (refs. 52, 53), *117*

P

Paddock, N. L., 228 (ref. 7), *237*
Padwa, A., 241 (ref. 10), 248 (ref. 10), *250*
Palkin, A. P., 124 (ref. 19), *139*
Panckhurst, D. J., 166 (ref. 15), *199*
Panontin, J. A., 161 (ref. 58), *164*
Parkinson, N., 6 (ref. 6), *25*

Pascal, P., 241 (ref. 8), 245 (ref. 33), *250*
Pasteur, L., 35, 38, *53*
Patterson, T. R., 14, *26*
Pauson, P. L., 78, 79 (ref. 1), 83, 97 (ref. 20), 98 (ref. 1), 106 (ref. 1), 112 (refs. 1, 20), 113 (ref. 20), *116*
Pavlik, F. J., 14 (ref. 28), *26*
Pavone, D., 222 (ref. 56), 223 (ref. 56), *225*
Pawlikowski, M. A., 11 (ref. 25a), *26*
Peacock, R. D., 127 (refs. 28, 30, 34), 128 (refs. 39, 41, 43, 45), 129, 132 (ref. 49), 138 (ref. 43), 139 (ref. 43), *139*, *140*
Peake, J. S., 182 (ref. 93), *201*
Pearson, R. G., 52 (ref. 89), *55*, *56*
Pechukas, A., 219–221 (ref. 45), *224*
Peppard, D. F., 42 (ref. 46a), *54*
Perl, H., 230 (ref. 12), 234 (ref. 12), *237*
Perlmutter-Hayman, B., 240 (ref. 4b), *249*
Perros, T. P., 15, 17, *26*
Perrotto, A., 17 (ref. 33), *26*
Peterhans, J., 113 (ref. 126), *119*
Peterson, D. T., 219, 220 (ref. 44), 222, *224*, *225*
Petru, F., 213 (ref. 24), 217 (ref. 24), 219 (ref. 24), *224*
Pfeiffer, P., 61, 65, 68, 72 (ref. 45), *74*, *75*
Pfitzner, H., 65 (ref. 26), *75*
Pflugmacher, A., 166 (ref. 14), 186 (ref. 107), 196 (refs. 107, 141), 198 (refs. 107, 141), *198*, *201*, *202*
Phillips, C. S. G., 166 (ref. 5), *198*
Pietrusza, E. W., 168 (ref. 24), 170 (ref. 24), 181 (ref. 24), 186 (ref. 24), *199*
Pietsch, G., 193 (ref. 137), 197 (ref. 137), *202*
Piper, T. S., 42, *54*
Pitzer, K. S., 111 (ref. 116), 112 (ref. 116), *119*
Plato, W., 50 (ref. 92), *55*
Plust, H. G., 181 (ref. 92), *201*
Podall, H. E., 22 (refs. 39, 40), *26*, 82, 96 (refs. 57, 65, 67, 68, 69a, 69b, 79, 80a), 97 (refs. 57, 65, 69b, 104), 103, 105, *116–119*
Pokras, L., 3 (ref. 5), 5, *25*
Polaroid Corp., 7 (ref. 19), *25*
Ponomarenko, V. A., 213, *224*

SUBJECT INDEX

A

Aldimine coordination compounds, 59–74
Anhydrous metal nitrates, 141–162
Asymmetric synthesis, 45–46
Azides of silane, 192

B

Barium hydride, 222
Beryllium nitrate, 154–156
Bromine nitrate, 162
Bromodisilanes, 193–195
Bromopolysilanes, 198
Bromosilanes, 182–183, 186–188

C

Calcium hydride, 219–222
Cesium hydride, 219
Chlorine nitrate, 161–162
Chlorodisilanes, 192–193
Chloropolysilanes, 197–198
Chlorosilanes, 180–182, 186–188
Chromium halide and oxyhalide complexes, 126–130, 134–135, 137
Chromium(III) nitrate, 159–160
Cobalt carbonyls, 100–101
Coordination compounds, aldimine and ketimine, 59–74
  optically active, 29–52
  polymeric, 1–25
Coordination polymers, 1–25
  plasticity of, 8–9
  requirements for linear, 10
  stability of, 7–8, 9
  synthetic methods, 11–25
Copper(II) nitrate, 154
Cyanide derivatives, of silane, 188–189
Cyano polymers, 2–3

D

1,3-Diketone coordination polymers, 11–15
Dinitrogen pentoxide, use in preparation of anhydrous nitrates, 156–162

Dinitrogen tetroxide, use in preparation of anhydrous nitrates, 143–156, 158–162
Disilane, halogen derivatives, 192–196
Dissymmetric synthesis, 45–46
Dry boxes, 91–93

E

Enantiomer production, by displacement, 46
  equilibrium methods, 44–45
Europium hydride, 222

F

Ferricyanides, 2–3
Ferrocyanides, 2–3
Fluorodisilanes, 192
Fluorosilanes, 178–180, 185–187

H

Halide complexes, of Ti, V, and Cr subgroup elements, 121–139
Halo polymers, 5–6
Halogen(I) alkoxides, 247–248
  carboxylates, 248–249
  fluorosulfonates, 246–247
  nitrates 161–162, 246
  oxides, 245–246
  perchlorates, 247
High pressure equipment, 84–90
Hydrides, apparatus for preparation of saline, 205–211
  dissociation pressures of saline, 204–205
  saline, 203–223
Hydroxides, 3–5
Hydroxo polymers, 3–5
8-Hydroxyquinoline coordination polymers, 15–17
Hypolfluorites, 241–245
Hypohalites, 239–245
  covalent, 241–245
  ionic, 241